Regenerative Medicine, Stem Cells and the Liver

Regenerative Medicine, Stem Cells and the Liver

Editor

David C. Hay

Principal Investigator

MRC Centre for Regenerative Medicine

University of Edinburgh

Edinburgh

UK

CRC Press

Taylor & Francis Group

Boca Raton London New York

CRC Press is an imprint of the
Taylor & Francis Group, an **informa** business

A SCIENCE PUBLISHERS BOOK

First published 2012 Science Publishers

Marketed and distributed by:

CRC Press
Taylor & Francis Group
6000 Broken Sound Parkway NW, Suite 300
Boca Raton, FL 33487-2742

ISBN-13: 978-1-57808-739-6 (hbk)

Visit the Taylor & Francis Web site at
http://www.taylorandfrancis.com

and the CRC Press Web site at
http://www.crcpress.com

Library of Congress Cataloging-in-Publication Data

Regenerative medicine : stem cells and the liver / editor, David C. Hay.
p. ; cm.
Includes bibliographical references and index.
ISBN 978-1-57808-739-6 (hardcover)
I. Hay, David C., 1973-
[DNLM: 1. Regenerative Medicine--methods. 2. Liver--growth & development. 3. Liver Regeneration. 4. Stem Cell Research. 5. Stem Cells--physiology. WO 515]

616.3'62--dc23

2012000718

Preface

The liver is an amazing organ which performs a myriad of functions necessary for life. The liver is highly resilient and the only internal human organ capable of large scale regeneration. During this process as little as 20–30% of healthy liver mass is all that is required to regenerate the whole liver. The regenerative capacity of the liver is captivating and some say this has been recognised for centuries, most notably in Greek mythology. The Titan Prometheus was chained to a rock and each day his liver was devoured by an eagle, only to grow back each night highlighting the liver's remarkable capacity for self-repair.

The primary regenerative response of the liver is provided by it's principal cell type, the hepatocyte. However when the liver is overwhelmed by insulting stimuli or chronically damaged, the hepatocyte response lost and a resident stem cell population is activated to regenerate functional tissue. Both acute and chronic liver injury can result in dramatic loss of human liver function and the onset of liver disease. The only real treatment for critically failing liver function is organ transplant. While transplantation is highly successful, the shortage of donor organs limits widespread implementation. Therefore researchers have been working together to explore scalable alternatives to support failing liver function. In particular, cell based therapies using human primary material have provided encouraging results. However, similar to organ transplant, there source of material is limited. This has led investigators to search for renewable cell sources, such as stem cells.

Stem cells are found in all multi-cellular organisms and divide producing identical copies of themselves, termed self renewal, or differentiate into specialized cell types. Stem cells are a very important source of cells which work to maintain organ function during human development, aging and injury. In addition to their role in tissue homeostasis, stem cells also offer great hope for developing novel medicines and cell based therapies to treat or replace diseased human tissue. Given the promise of stem cell biology and the potential it offers to human medicine in the 21st Century we have written this book.

We begin by outlining the complex nature of human liver disease and examining the potentials that stem cell based approaches have to offer.

We follow this by highlighting potential sources of cell populations for therapy and discuss the challenge with working with primary human material. All too frequently news headlines focus on the benefits of transplanting stem cell based products in patients. However in reality one of the most far-reaching benefits of stem cell research to society will be the accelerated development of novel, safer and more effective human medicines. Therefore the focus of the four following chapters discusses the role that stem cell technology has to play in understanding human biology, human disease and its application as a model to improve pharmaceutical drug development and safety. Following on from this we examine the important role that cell banking and manufacturing has to play in the scale up, quality assurance and clinical grade manufacture of stem cell derived products. This is followed by three chapters which focus on the use of cell based therapies for the treatment of human liver disease. We conclude the book with a chapter devoted to the challenges, technically and ethically, that exist for stem cell derived products for research and therapy.

I have been very lucky to work with a number of the world's experts on this project and we have endeavoured to provide comprehensive coverage of the field. I hope you enjoy reading the book.

Dr. David C. Hay
University of Edinburgh March 2012

Contents

List of Contributors

Brian Adams
Department of Genetics, Yale Stem Cell Center, Yale School of Medicine, 10 Amistad, New Haven, CT. 06520, USA.
Email: brain.adams@yale.edu

Richard A. Anderson
MRC Centre for Regenerative Medicine, University of Edinburgh, 49 Little France Crescent, Edinburgh, EH16 4SB, U.K.
Email: r.anderson@hrsu.mrc.ac.uk

Petter Björquist
Cellectis Stem Cells Cellartis AB, Arvid Wallgrens Backe 20, 41346 Göteborg, Sweden.
Email: petter.bjorquist@cellectis.com

James R. Black
MRC Centre for Regenerative Medicine, University of Edinburgh, 49 Little France Crescent, Edinburgh, EH16 4SB, U.K.
Email: jblack2@staffmail.ed.ac.uk

Luke Boulter
Research Fellow, Queens Medical Research Institute, 47 Little France Crescent, Edinburgh, EH16 4TJ, UK.
Email: lboulter@staffmail.ed.ac.uk

Nicholas James Davies
Centre for Liver Research and NIHR Biomedical Research Unit, University of Birmingham,Birmingham, U.K.
Email: n.j.davies@bham.ac.uk

Paul A. De Sousa
Scottish Centre for Regenerative Medicine, University of Edinburgh, Edinburgh, EH16 4SB, UK.
Email: paul.desousa@ed.ac.uk

Mia Emgård
Cellectis Stem Cells Cellartis AB, Arvid Wallgrens Backe 20, 41346 Göteborg, Sweden.
Email: mia.emgard@cellectis.com

Nicholas R.F. Hannan
Anne McLAren Laboratories for Regenerative Medicine, Department of Surgery, University of Cambridge, UK.
Email: nrfh2@cam.ac.uk

Sian E. Harding
Professor of Cardiac Pharmacology, National Heart and Lung Institute, Faculty of Medicine, Imperial College London, UK, London SW7 2AZ, UK, Member of the Nuffield Council on Bioethics.
Email: sian.harding@imperial.ac.uk

David C. Hay
MRC Centre for Regenerative Medicine, University of Edinburgh, 49 Little France Crescent, Edinburgh, EH16 4SB, U.K.
Email: davehay@talktalk.net

Humphrey Hodgson
Royal Free Campus, UCL Medical School, London NW3 2PF.
Email: hhodgson@medsch.ucl.ac.uk

Eriona Hysolli
Department of Genetics, Yale Stem Cell Center, Yale School of Medicine, 10 Amistad, New Haven, CT. 06520, USA.
Email: eriona.hysolli@yale.edu

John P. Iredale
MRC Centre for Regenerative Medicine, University of Edinburgh, 49 Little France Crescent, Edinburgh, EH16 4SB, U.K.
Email: john.iredale@ed.ac.uk

Ji-Hyun Kim
Department of Genetics, Yale Stem Cell Center, Yale School of Medicine, 10 Amistad, New Haven, CT. 06520, USA.
Email: ji-hyun.kim@yale.edu

Janet W.C. Kung
MRC Centre for Regenerative Medicine, University of Edinburgh, 49 Little France Crescent, Edinburgh, EH16 4SB, U.K.
Email: janetkung@gmail.com

Itziar de Lecuona
Lecturer, Dept. of Public Health, School of Medicine, Bioethics and Law Observatory, Barcelona Science Park, University of Barcelona, 08028 Barcelona, Spain.
Email: itziardelecuona@ub.edu

Renjing Liu
Department of Genetics, Yale Stem Cell Center, Yale School of Medicine, 10 Amistad, New Haven, CT. 06520, USA.
Email: renjing.liu@yale.edu

Scott McRae
Cell Guidance Systems, Babraham Research and Control-HPA Campus, Cambridge, UK.
Email: scott.mcrae@cellgs.com

Claire N. Medine
MRC Centre for Regenerative Medicine, University of Edinburgh, 49 Little France Crescent, Edinburgh, EH16 4SB, U.K.
Email: cmedine@staffmail.ed.ac.uk

Philip Noel Newsome
Centre for Liver Research and NIHR Biomedical Research Unit, University of Birmingham, Birmingham, UK.
Email: p.n.newsome@bham.ac.uk

In-Hyun Park
Department of Genetics, Yale Stem Cell Center, Yale School of Medicine, 10 Amistad, New Haven, CT. 06520, USA.
Email: inhyun.park@yale.edu

Catherine M. Payne
MRC Centre for Regenerative Medicine, University of Edinburgh, 49 Little France Crescent, Edinburgh, EH16 4SB, U.K.
Email: cateduff@btinternet.com

Prakash Ramachandran
MRC Centre for Inflammation Research, University of Edinburgh, 47 Little France Crescent, Edinburgh, EH16 4TJ, U.K.
Email: pramacha@staffmail.ed.ac.uk

S. Tamir Rashid
Anne McLAren Laboratories for Regenerative Medicine, Department of Surgery, University of Cambridge, UK.
Email: str29@cam.ac.uk

James A. Ross
MRC Centre for Regenerative Medicine, University of Edinburgh, 49 Little France Crescent, Edinburgh, EH16 4SB, UK.
Email: j.a.ross@ed.ac.uk

Clare Selden
Royal Free Campus, UCL Medical School, London NW3 2PF.
Email: cselden@medsch.ucl.ac.uk

Glyn Stacey
National Institute for Biological Standards-HPA, Blanche Lane, South Mimms, Herts, EN6 3QG.
Email: glyn.stacey@nibsc.hpa.org.uk

Gareth Sullivan
MRC Centre for Regenerative Medicine, University of Edinburgh, 49 Little France Crescent, Edinburgh, EH16 4SB, UK.
Email: garethsully@hotmail.com

Amy Thomas
Royal Free Campus, UCL Medical School, London NW3 2PF.
Email: athomas@medsch.ucl.ac.uk

Jordi L.Tremoleda
Investigator, MRC Centre for Clinical Sciences, Faculty of Medicine, Imperial College London, London W12 0NN, UK.
Email: Jordi.lopez-tremoleda@imperial.ac.uk

Ludovic Vallier
Anne McLAren Laboratories for Regenerative Medicine, Department of Surgery, University of Cambridge, UK.
Email: lv225@cam.ac.uk

Xiao-Ling Zhou
Department of Genetics, Yale Stem Cell Center, Yale School of Medicine, 10 Amistad, New Haven, CT., 06520, USA.
Email: xiaoling.zhou@yale.edu

1

Stem Cell Therapy in the Context of Chronic Liver Disease: Promise and Pitfalls

***Prakash Ramachandran*[1,a] and *John P. Iredale*[1,b,*]**

Regenerative medicine occupies a unique position in research and translation, drawing on the disparate but overlapping fields of developmental biology, cell biology, genetics, epigenetics, inflammation and of course stem cell biology. Key attributes of each of these fields are critical when organ or tissue restitution occurs with repair that restores a normal architecture and replacement of the component cell types that maintain structure and specific function. Nowhere is this more important or arguably better exemplified than in liver disease.

To date the major interest in the deployment of stem cells in the context of liver disease (and numerically the greatest number of publications) has been focused on the area of parenchymal restitution. Restoration of the liver parenchyma requires the replacement of diseased or failing hepatocytes with new hepatocytes, either exogenously derived from stem or progenitor cell elements or, theoretically, derived from stimuli applied to the liver progenitor cell component. Is this approach likely to be successful in the immediate future as a new approach to restoring normal hepatic function to diseased livers? The single biggest therapeutic challenge clinically in the context of chronic liver disease is fibrosis and its end stage cirrhosis

[1]MRC Centre for Inflammation Research, University of Edinburgh, 47 Little France Crescent, Edinburgh, EH16 4TJ, U.K.
[a]Email: pramacha@staffmail.ed.ac.uk
[b]Email: john.iredale@ed.ac.uk
*Corresponding author

with attendant hepatocellular failure. Fibrosis and cirrhosis result from chronic inflammatory injury within the liver and represent the final common pathway of virtually all chronic liver diseases from paediatric metabolic disorders to alcohol excess. Our view is that current evidence suggests that the approach of "adding back" hepatocytes is unlikely to be uniformly successful unless the underlying inflammatory liver disease and in particular the consequent fibrosis, which characterizes all chronic liver diseases, can be brought under control.

Virtually all chronic liver diseases are associated with inflammation to a greater or lesser extent. This is manifestly obvious in acute viral hepatitis or aggressive autoimmune disease. The presence of chronic inflammation in turn leads to the activation of myofibroblasts, the recruitment of key inflammatory cell elements and the development of fibrosis. Initially perisinusoidal, fibrosis eventually extends to link the vascular structures and distort the liver anatomy. At this stage, if there is regenerative activity from hepatocytes or progenitor cells these in turn form spheres of poorly functioning liver parenchyma and the resulting structure is termed cirrhosis.

Thus, in the context of chronic liver disease, hepatocyte dysfunction may result from direct damage from the causative stimulus, as a result of intense inflammatory activity or due to changes in specific function mediated by the abnormal pericellular milieu and altered cell-matrix interactions. The latter may be important not only for hepatocellular function but also hepatocyte and hepatic progenitor cell proliferation (Kallis et al. 2011).

The delivery of new hepatocytes derived from progenitor or stem cells in this context might not be expected to result in a dramatic return of organ function. Indeed any new hepatocytes would be susceptible to the injurious agent directly and the consequences of the altered milieu. Additionally, phenotypic changes in the sinusoidal endothelium following chronic injury may also impact on the recruitment and engraftment of exogenous hepatocytes. Put another way, planting the seed in poor quality soil is unlikely to deliver a significant harvest.

That is not to say, however, that these problems are insurmountable. It is perhaps not surprising that the greatest successes in hepatocyte transplantation (in many ways an analogous experiment differing only in origin of the transplanted cells) have been demonstrated in the context of orphan paediatric metabolic disorders, such as Crigler–Najar, Primary Hyperoxaluria or Urea cycle disorders (Meyburg et al. 2009). In such conditions the metabolic defects are predominantly expressed in the liver but the organ damage results elsewhere, whilst the liver parenchyma remains normal, with little by way of liver inflammation and fibrosis. In these genetic conditions, endogenous regenerative systems will be subject to the same metabolic defects. Therefore, the transplanted, genetically normal, hepatocytes will have a positive selection pressure and can induce a good functional recovery, acting as either a bridge to liver transplantation or indeed a medium term therapeutic solution (Meyburg et al. 2009). Similar findings are seen dramatically in animal models in which transplanted stem cell derived hepatocytes may have a growth and survival advantage over the metabolically compromised host cells and in which the engrafted cells proliferate and deliver a major proportion of the metabolic function of the liver's hepatocytes (Shafritz and Oertel 2011, Rhim et al. 1994).

Moreover, the delivery of cells to a heterotypic site, for example the peritoneum, might overcome some of the problems of the specific liver milieu (Baldini et al. 2008). Furthermore, it is not inconceivable that an uninjured peritoneal surface could provide anchorage and cell-cell and cell matrix signals that may preserve and foster hepatocyte function. An alternative heterotypic site might, for example be the spleen (Payne et al. 2011). Additionally, it may be possible to engineer cells that are resistant to the injurious agent thus eliminating direct damage to engrafted cells. However, assuming that those cells also shared other attributes of the normal cell phenotype for hepatocytes, the adverse inflammatory and fibrotic milieu might still result in changes to function.

An alternative approach utilizing stem cells which may come to the forefront in the immediate future is the development of a bioartifical, or liver support device. This is likely to be particularly valuable in the clinical context of acute liver failure, where massive hepatocellular necrosis can result in previously well patients developing rapidly progressive liver and multi-organ failure. Often liver transplantation is the only curative option in this setting, and due to the lack of available organs other treatments are badly needed. Clearly the liver milieu in this setting would not be conducive to cellular transplantation. Previous attempts to use liver assisting devices, akin to renal replacement therapy, as a bridge to transplantation or to hepatic recovery have not demonstrated a survival benefit (Liu et al. 2004). Principally, the complexity of the biological functions performed by the liver mean that standard liver dialysis systems cannot compensate for this loss. Cellular systems have also been tried to overcome this (Carpentier et al. 2009). However, the sheer volume of metabolically active human hepatocytes required for an effective liver device has, to date, significantly impacted on developments in this field. One enormous advantage provided by stem cell derived hepatocytes is that they can be manufactured to a uniform high standard, theoretically in infinite numbers, giving exciting opportunities for their use in "off the shelf" bioartificial liver support devices.

The inflamed and fibrotic liver is, also an excellent model for matrix remodelling, parenchymal restitution (from hepatocytes and progenitor cell elements) and a return of normal or more normal architecture, if the inflammatory and fibrotic processes can be terminated. Previously considered irreversible, there is now a wealth of evidence from both animal and human models that with cessation of the injurious stimulus, liver fibrosis will undergo remodelling with changes in the matrix from a fibrotic to a more normal sinusoidal pattern (degradations of collagens I and III and a return to a matrix rich in laminins and collagen IV) (Iredale 2007). This will result not only in return to a more normal architecture, but is also associated with hepatocyte proliferation and a hepatic progenitor cell response (Kallis et al. 2011).

During the last 10 years the events underpinning this process of fibrosis remodelling and resolution have become clear. Whereas in progressive fibrogenesis myofibroblasts become activated and express high levels of fibrillar collagens I and III, and concurrently high levels of the tissue inhibitors of metalloproteinases TIMPs 1 and 2, following withdrawal of the injurious stimulus this pattern reverses and the expression of TIMPs 1 and 2 and the fibrillar collagens drops rapidly to become undetectable. Moreover, the myofibroblasts which populate the liver scar undergo apoptosis, thereby withdrawing both the source of the new fibrillar matrix and the source of the proteins which protect the matrix from degradation (Iredale et al. 1998). The TIMPs are very potent metalloproteinase inhibitors, binding in a non-covalent manner and acting against all members of the MMP family. Thus, during progressive fibrosis, fibrillar matrix accumulates not only as a result of enhanced synthesis but also as a result of a failure of matrix degradation. Following withdrawal of the injurious stimulus and the drop in TIMP levels MMP mediated matrix degrading activity, previously held in check, can occur and the remodelling process commences (Iredale 2007, Ramachandran and Iredale 2009).

Central to the process of matrix remodelling is the macrophage. It has been known for many years that macrophages are necessary to drive the inflammation and fibrogenic process during progressive fibrosis (Karlmark et al. 2009). However, an interesting and apparently counterintuitive observation—that the maximum number of macrophages within a liver during experimental fibrosis resolution occurs as matrix degradation commences, suggests a pro resolution function also. In combination with the observation that the major TIMP expressing cell, the myofibroblast, undergoes apoptosis during resolution of fibrosis has led to the tantalizing suggestion that macrophages may be critical for fibrosis resolution. We and others have now demonstrated unequivocally, using models in which macrophages can be selectively depleted, that macrophages are necessary for matrix degradation in the spontaneous resolution of liver fibrosis (Duffield et al. 2005). Moreover MMPs derived from those macrophages including a key collagenase (MMP-13) and elastase (MMP-12) in addition to promiscuous enzymes that will degrade the fibrillar collagens following initial collagenase mediated cleavage (MMPs 9 and 7) are almost exclusively derived from macrophages in experimental rodent models (Fallowfield et al. 2007). Intriguingly, the macrophages that we have observed infiltrating the liver and fulfilling the matrix degrading functions also express the mitogens TWEAK and IGF1 both of which have now been identified as important for hepatic progenitor cell proliferation and hepatocyte proliferation respectively.

These observations highlight the tantalizing possibility that macrophage therapy may be a valuable approach to treating the underlying inflammatory and fibrotic condition within the liver, in addition to promoting regeneration. Moreover, in ground breaking work, the Sakaida group demonstrated clear evidence that infusion of bone marrow elements was associated with enhanced matrix degradation and resolution of fibrosis and that cells secreting the MMP repertoire of macrophages were likely mediating this effect (Sakaida et al. 2004). Our group and others have subsequently demonstrated that myeloid elements, particularly macrophages, are the cells that are essential to fulfilling this function in experimental engraftment and in addition can enhance the hepatic progenitor cell response in this context (Thomas et al. 2011). Furthermore, in work using renal models of inflammation and fibrosis it has been demonstrated that macrophages altered to over-express IL-10 or Il-4 will function as a potent anti-inflammatory agent (Wilson et al. 2002, Kluth et al. 2001). Indeed, because the macrophages will be home to an area of inflammation they can in some respect be considered a targetted therapy in these models. Macrophages are of course in turn derived from myeloid and bone marrow stem cell and progenitor elements. Thus at an early stage it is likely that we will see stem cell related trials, in this case of macrophage therapy, being deployed to treat liver fibrosis.

In conclusion, there remains enormous promise for the use of regenerative medicine technology in the clinical treatment of human liver disease. Hepatocyte transplantation and the development of bioartificial liver devices are likely to find their niche in the management of metabolic liver disease and acute liver failure respectively. In the context of chronic disease, the liver remains an extraordinary paradigm for the study of inflammation, fibrosis resolution and regeneration. Stem cell derived therapies are likely ultimately to impact on parenchymal reconstitution but will be more effective if they can be concurrently delivered with an antifibrotic therapy to improve the milieu into which they will engraft. Macrophage therapy for fibrosis, in turn is the result of expansion of progenitor and stem cell elements derived from the bone marrow, shows enormous promise as a targetted cell therapy of fibrosis *in vivo*.

References

Baldini E, R Cursio, G De Sousa et al. 2008. Peritoneal implantation of cryopreserved encapsulated porcine hepatocytes in rats without immunosuppression: viability and function. Transplant Proc. 40(6): 2049–52.

Carpentier B, A Gautier and C Legallais. 2009. Artificial and bioartificial liver devices: present and future. Gut. 58(12): 1690–702.

Duffield JS, SJ Forbes, CM Constandinou et al. 2005. Selective depletion of macrophages reveals distinct, opposing roles during liver injury and repair. J Clin Invest. 115(1): 56–65.

Fallowfield JA, M Mizuno, TJ Kendall et al. 2007. Scar-associated macrophages are a major source of hepatic matrix metalloproteinase-13 and facilitate the resolution of murine hepatic fibrosis. J Immunol. 178(8): 5288–95.

Iredale JP. 2007. Models of liver fibrosis: exploring the dynamic nature of inflammation and repair in a solid organ. J Clin Invest. 117(3): 539–48.

Iredale JP, RC Benyon, J Pickering et al. 1998. Mechanisms of spontaneous resolution of rat liver fibrosis. Hepatic stellate cell apoptosis and reduced hepatic expression of metalloproteinase inhibitors. J Clin Invest. 102(3): 538–49.

Kallis YN, AJ Robson, JA Fallowfield et al. 2011. Remodelling of extracellular matrix is a requirement for the hepatic progenitor cell response. Gut. 60(4): 525–33.

Karlmark KR, R Weiskirchen, HW Zimmermann et al. 2009. Hepatic recruitment of the inflammatory Gr1+ monocyte subset upon liver injury promotes hepatic fibrosis. Hepatology. 50(1): 261–74.

Kluth DC, CV Ainslie, WP Pearce et al. 2001. Macrophages transfected with adenovirus to express IL-4 reduce inflammation in experimental glomerulonephritis. J Immunol. 166(7): 4728–36.

Liu JP, LL Gluud, B Als-Nielsen et al. 2004. Artificial and bioartificial support systems for liver failure. Cochrane Database Syst Rev. (1): CD003628.

Meyburg J, J Schmidt and GF Hoffmann. 2009. Liver cell transplantation in children. Clin Transplant. Suppl. 21: 75–82.

Payne CM, K Samuel, A Pryde et al. 2011. Persistence of functional hepatocyte-like cells in immune-compromised mice. Liver Int. 31(2): 254–62.

Ramachandran P and JP Iredale. 2009 Reversibility of liver fibrosis. Ann Hepatol. 8(4): 283–91.

Rhim JA, EP Sandgren, JL Degen et al. 1994. Replacement of diseased mouse liver by hepatic cell transplantation. Science. 263(5150): 1149–52.

Sakaida I, S Terai, N Yamamoto et al. 2004. Transplantation of bone marrow cells reduces CCl4-induced liver fibrosis in mice. Hepatology. 40(6): 1304–11.

Shafritz DA and M Oertel. 2011. Model systems and experimental conditions that lead to effective repopulation of the liver by transplanted cells. Int J Biochem Cell Biol. 43(2): 198–213.

Thomas JA, C Pope, D Wojtacha et al. 2011. Macrophage therapy for murine liver fibrosis recruits host effector cells improving fibrosis, regeneration, and function. Hepatology. 53(6): 2003–15.

Wilson HM, KN Stewart, PA Brown et al. 2002. Bone-marrow-derived macrophages genetically modified to produce IL-10 reduce injury in experimental glomerulonephritis. Mol Ther. 6(6): 710–7.

2

Sources of Human Liver Cells and the Challenge of Working with Primary Tissue

Janet W.C. Kung[1,a,*] and *James A. Ross*[1]

Introduction

Liver cirrhosis represents the final common histologic pathway for a wide variety of chronic liver diseases. With the rising incidence of alcoholic liver disease, hepatitis C and, more recently, non-alcoholic liver disease associated with obesity, cirrhosis places an increasing burden on healthcare worldwide. Currently the only curative treatment is liver transplantation. While liver transplant has a relatively good 5-yr survival, donor organ shortage costs many lives every year; and despite improvements in donor schemes, the use of broader donor criteria, and advances in surgical techniques, the increase in the number of transplants has been modest (van der Meulen et al. 2007). New figures from the NHS Blood and Transplant survey reveal that in the 12-mon period to June 2011, the number of people waiting for a new liver in the UK rose by 25% when compared with the 12 mon to June 2010 (NHS Blood and Transplant [Internet] 2011). Alternative treatments are thus urgently needed.

The generation of functional hepatocytes is of great therapeutic interest with applications in drug screening and disease modelling, human bioartificial liver construction, and, potentially, in hepatocyte transplantation. The ability to use human cell types in the pharmaceutical

[1]MRC Centre for Regenerative Medicine, University of Edinburgh, 49 Little France Crescent, Edinburgh, EH16 4SB.
[a]Email: janetkung@gmail.com
*Corresponding author

industry not only allows expedition of novel human drug development, but also takes into account variability in drug metabolism because of cytochrome P450 polymorphisms. Phase I trials employing a bio-artificial liver device have shown some promise as patients exhibited improvement in neurological state and haemodynamics, but this approach is currently limited by the use of xenogeneic materials. Such bioartificial devices permit some restoration of liver metabolic function and are, in the future, likely to act as a bridge to liver transplantation in patients with advanced or end-stage liver failure (Gerlach et al. 2008, McKenzie et al. 2008). Attempts at hepatocyte transplantation to treat some liver-based inborn errors of metabolism and liver failure (Fisher and Strom 2006, Puppi and Dhawan 2009, Smets et al. 2008) as an alternative to orthotopic liver transplantation have shown transient clinical improvement and/or partial correction of the underlying metabolic defect. Whilst significant progress has been made to develop the technique, there are many obstacles to successful hepatocyte transplantation and the main constraint remains the difficulty in sourcing and maintaining viable hepatocytes. This chapter aims to outline the sources of human hepatocytes and discuss the challenges faced by researchers working with primary human tissue.

Sources of human hepatocytes

Mature (adult) hepatocytes

The human liver has a remarkable regenerative capacity. Following acute liver injury such as partial hepatectomy, the tissue mass is restored by mitotic division of mature hepatocytes (Fausto et al. 2006). However, hepatocytes in culture rapidly lose their *in vivo* phenotypic characteristics and functional abilities with a decline in cytochrome P450 levels occurring within hours of isolation (Elaut et al. 2006). The compound effect of endotoxin-containing collagenase digestion, disruption of normal tissue architecture (cell-cell and cell-matrix interactions), and ischaemic-reperfusion injury during the isolation process triggers a cascade of signalling pathways leading to hepatocyte apoptosis and anoikis (Elaut et al. 2006). Despite decades of research, it remains difficult to maintain liver-specific functions of primary hepatocytes in culture for more than a week without an adequate supportive microenvironment either in the form of a three-dimensional extracellular matrix or the presence of xenogeneic feeder layers. Nonetheless, de-differentiation of primary hepatocytes appears to be an inevitable eventuality. Attempts at immortalization of primary hepatocytes are currently only partially successful as proliferation and hepatocellular function can appear mutually exclusive (van de Kerkhove et al. 2005). In addition, hepatocytes do not spontaneously redifferentiate after *in vitro* culture. This difficulty with maintaining expandable functional hepatocytes, together with the availability of adequate liver tissue, the functionality of cryopreserved cells and the need for recipient immunosuppression, limits their use in hepatocyte transplantation. Furthermore, a careful selection of target patient populations will need to be exercized as transplantation of mature hepatocytes into patients with metabolic liver disease only results in partial correction of function whilst cirrhotic patients receive little or no benefit as scar reduction and tissue remodelling are likely to be critical (Kung and Forbes 2009).

Aside from hepatocytes, the liver has a population of facultative hepatic progenitor cells (in rodents called oval cells) that are recruited for regeneration when hepatocyte mitosis is overwhelmed during massive or chronic injury. The mechanisms controlling progenitor activation are under intense investigation, and while a number of inflammatory cytokines have previously been implicated in rodent models, recent studies have indicated an important role for Hedgehog signalling in progenitor activation in alcoholic steatohepatitis in mice and humans (Jung et al. 2008, Fleig et al. 2007). Observations of the gradient formed by young hepatocytes with higher proliferation potential in periportal areas and more mature hepatocytes in pericentral areas have led to the "streaming liver" hypothesis, which suggests that normal liver turnover is similar to intestinal regeneration (Barker et al. 2007, Sato et al. 2011), with periportal progenitor cells proliferating extensively, and upon migration towards the central vein, differentiating into mature hepatocytes and cholangiocytes (Zajicek et al. 1985). This hypothesis is supported by a recent study wherein clonal proliferative units were traced in human livers (Fellous et al. 2009). Accordingly, hepatic progenitors are thought to be located in Canals of Hering in the adult liver and possibly in the ductal plate in the foetal liver. Protocols for the isolation, culture and propagation of hepatic progenitors in rodent models of liver disease (prior exposure to chemicals, carcinogens, dietary manipulation or liver injury) have been established for several decades, however, the relevance of these cells as a source of transplantable cells is not clear. It is not until recently that progenitor cells from unmanipulated rodent livers have been isolated and shown to be able to proliferate extensively in culture and give rise to hepatocytes after engraftment into the liver (Wright et al. 2008, Sahin et al. 2008, Chen et al. 2010). Although it remains questionable whether progenitors in rodents and humans represent cells of comparable functional phenotype, inferences made from rodent studies could be extrapolated to facilitate production of clinically usable material. To date, only one research group has described isolation of liver progenitors from human adult livers (Schmelzer et al. 2006). The paucity of such studies may in part be due to tissue availability as competition for organs exists between hepatocyte and liver transplantation, but also donor organs that are deemed unsuitable for whole organ transplantation which are thence released for research or hepatocyte transplantation are often of marginal quality. As an alternative to hepatocyte transplantation, the adult liver progenitor cell pool might be selectively targeted via pharmacological means to manipulate their proliferation capacity. However, only the inhibition of proliferation has yet been shown. In a mouse model of chronic liver injury, imatinib mesylate inhibited the oval cell response and early fibrinogenesis by inhibition of c-kit activity (Knight et al. 2008). Tyrosine kinase inhibitors may therefore be suited for the treatment of chronic liver disease and potentially as a preventative measure in hepatic carcinogenesis.

Foetal hepatic progenitor cells

A number of studies show that foetal liver progenitors might be superior as sources of hepatocyte transplantation. Compared with mature hepatocytes, foetal progenitors have the advantage of being (i) bipotential, capable of regenerating hepatocytes and the biliary system; (ii) highly proliferative; (iii) less apoptotic following transplantation; (iv) less immunogenic; and (v) smaller in size, allowing easier and safer intraportal injection and dispersion (Oertel et al. 2006). The use of foetal liver progenitor cells also abrogates the issue of forced differentiation of pluripotent stem cells as foetal progenitors have undergone sufficient morphological and physiological differentiation so that they are committed to a hepatic fate and yet they retain their "stemness".

The phenotype of foetal human liver progenitor cells remains controversial. A range of cell markers based on rodent studies, such as Liv2 (Watanabe et al. 2002, Takashimizu et al. 2009), E-cadherin (Nitou et al. 2002) and delta like kinase-1 (Dlk-1) (Tanimizu et al. 2003), have only been characterized in human livers by immunodetection methods *in vitro* (Inada et al. 2008, Terrace et al. 2007). To date, the only convincing evidence to show that liver progenitors can be isolated from human foetal livers comes from immunoselection for epithelial cell adhesion molecule (EpCAM)-positive cells (Schmelzer et al. 2007). *In situ* studies reveal that $EpCAM^+$ foetal liver progenitors are located in the ductal plate. Once isolated, these cells are capable of self-renewal and clonogenic expansion, as well as differentiation into both hepatocytic and biliary lineages in defined culture conditions (Schmelzer et al. 2007). Moreover, purified $EpCAM^+$ foetal liver progenitors when transplanted are able to engraft the livers of immunodeficient adult mice yielding mature human liver tissue (Schmelzer et al. 2007). Although this represents a major advance towards developing protocols that can be translated to clinically relevant therapy, the activating signal for progenitor cell proliferation and differentiation as well as the optimal host liver environment to allow foetal hepatoblasts to behave as hepatic progenitors remain to be fully elucidated. Further, issues concerning cryopreservation, differences in age of donor/host cells and species differences, all of which affecting repopulation and maturation, would need to be addressed as xenotransplantation studies have shown inconsistent results. Transplantation of cryopreserved early foetal hepatoblasts (11–13 wk of gestation) into newborn athymic mouse livers resulted in up to 10% liver repopulation and cells expressing adult hepatocyte markers (Mahieu-Caputo et al. 2004), while foetal hepatoblasts (14–17 wk of gestation) transplanted into immunodeficient heterozygous albumin-urokinase-type plasminogen activator mice resulted in a significantly lower degree of liver repopulation than adult human hepatocytes (Haridass et al. 2009).

Another potential stem cell population, side population (SP) cells, has been found to contribute to haematopoietic and epithelial lineages in the early gestational phase of human liver development (Terrace et al. 2009). SP cells have been isolated using fluorescence-activated cell sorting based on their ability to efflux DNA-labelling Hoechst dye (Goodell et al. 2005), a phenotype determined by expression of ATP-binding cassette (ABC) transporters encoded by the multi-drug resistance (MDR)-1 gene (Terrace et al. 2009). Their location *in situ*, however, remains uncertain, not least because of the widespread distribution of ABC transporters in the liver (Plosch et al. 2005); clearly the vast majority of cells in the liver expressing ABC proteins are not stem cells (Forbes and Alison 2006).

Bone marrow stem cells

There was great interest in the possibility of the so-called adult stem cell plasticity when Y-chromosome-positive hepatocytes were identified in female patients receiving a male bone marrow (BM) graft. It was suggested that extrahepatic cells may generate hepatocytes *in vivo*. Further demonstration of BM transplant rescuing a mouse model of tyrosinaemia (Lagasse et al. 2000), a hereditary defect of a hepatocytic enzyme, led to intense investigation of BM stem cells as a potential source of liver stem cells and as a means to regenerate the cirrhotic liver. Subsequent analysis revealed that the apparent plasticity is due to fusion with host hepatocytes under unique selective pressure rather than transdifferentiation of extrahepatic cells. The correction of the metabolic defect seen in murine models was therefore secondary to fusion-mediated gene transfer from wild-type BM cells to the defective hepatocytes. To date, the mechanism of actions of BM stem cells within the human liver remains elusive. However, there is recognition that irrespective of mechanism, the number of hepatocytes produced is generally too low to have any clinical impact at present (Houlihan and Newsome 2008). Rather than transdifferentiation to hepatocytes, BM stem cells might have a role in remodelling the fibrosis seen in advanced liver disease. Whilst reduction of fibrosis was observed in rodent models of chronic liver injury after BM infusion, a process mediated by increased matrix metalloproteinase-9 secretion (Sun et al. 2008, Oyagi et al. 2006), there is a need for caution as liver fibrinogenic cells (hepatic stellate cells and myofibroblasts) can also originate from BM stem cells (Russo et al. 2006, Kisseleva et al. 2006). There are therefore studies underway to identify the best BM cells for matrix degradation and stimulation of liver regeneration whilst avoiding the cell populations that can give rise to hepatic scarring.

There are an accumulating number of published human clinical studies investigating the effects of BM stem cell therapy in patients with liver disease. A recent randomized controlled trial involving 24 patients with alcoholic cirrhosis showed that granulocyte-colony stimulating factor (G-CSF) mobilizes intrahepatic progenitor cells, and induced liver regeneration (Spahr et al. 2008). The study was too small to detect a survival effect but was nonetheless encouraging. Autologous BM mononuclear cell transplantation also showed encouraging preliminary results in patients with advanced liver cirrhosis. Thirty patients on the liver transplant waiting list received BM cell therapy or placebo in a non-blinded fashion. Patients in the treatment arm of the study demonstrated an increase in albumin level and a decrease in their Child-Pugh score at the study endpoint compared with controls (Lyra 2007). Smaller observational clinical studies illustrated the safety and feasibility of G-CSF administration followed by leukapharesis and reinfusion of CD34-positive cells in patients with chronic liver disease, who derived some benefits from this treatment regime showing improved clinical status and biochemistry (Levicar et al. 2008, Yannaki et al. 2006). BM stem cell therapy is an attractive option for the treatment of liver cirrhosis as we already manipulate and transplant these cells in the clinic. However, it is paramount to conduct larger well-designed clinical trials to fully establish the safety profile of such therapies, as well as to define the target patient groups with treatment efficacy assessed by standardized protocols.

Adipose tissue-derived stem cells

Mesenchymal stem cells (MSCs) can be derived from various adult tissues and can be differentiated *in vitro* into many cell types. It is encouraging to note that there are currently over 60 MSC ongoing clinical trials with no safety risk demonstrated in humans so far (Newman et al. 2009). Human adipose tissue-derived MSCs (AT-MSCs) can be differentiated to become hepatocyte-like, with gene expression, morphology and metabolic activity similar to hepatocytes (Sgodda et al. 2007, Banas et al. 2007). In a xenogeneic transplantation model of liver regeneration, long term engraftment of AT-MSCs-derived hepatocyte-like cells was demonstrated and human albumin levels were measured at 15 times higher in the serum of mice transplanted with pre-differentiated AT-MSCs compared with undifferentiated AT-MSCs at 10 wk post-transplantation (Aurich et al. 2009). Further, a repopulation rate of more than 10% greatly exceeds the accepted repopulation threshold of 1–5% deemed sufficient to correct liver disorders in most injury models (Aurich et al. 2009), making AT-MSCs a potential candidate and readily available source for the generation of adult stem cell based liver therapies. A key factor will be some demonstration of the long-term phenotypic stability of the transplanted cells and their ability to contribute to tissue homeostasis.

Embryonic stem cells and induced pluripotent stem cells

Pluripotent human embryonic stem cells (ESCs) can be differentiated towards the hepatic lineage both by spontaneous differentiation via the formation of embryoid bodies and by directed differentiation using inducing factors toward ESC-derived hepatic endoderm (reviewed in (Asahina et al. 2006, Dalgetty et al. 2009)). Cells generated in this fashion are hepatocyte-like and can engraft with both mature hepatocytes and cholangiocytes when transplanted although the level of liver repopulation remains very low (Gouon-Evans et al. 2006, Heo et al. 2006). Continuing efforts are being made to improve the ability to derive, purify and scale ESC-derived hepatocyte-like cells *in vitro*, and exposure of ESCs to Wnt3a signalling was recently found to improve hepatocyte functionality both *in vitro* and *in vivo* (Hay et al. 2008). Hepatocyte-like cells can now be produced in sufficient quantities for both *in vitro* and *in vivo* experimental applications essential to clinical translation. For therapeutic utility, however, such cells should be immune-matched to obviate the need for immunosuppression. Furthermore, ethical concerns and the possibility of teratoma formation may constitute limitations to the use of ESC-derived hepatocytes. The development of ESC-like induced pluripotent stem cells (iPSCs) may therefore prove useful. Using Oct3/4, Sox2, cMyc and Klf4 transfection, iPSCs can be generated from terminally-differentiated adult somatic cells (Takahashi et al. 2007, Yamanaka 2007), potentially circumventing problems arising due to human leukocyte antigen mismatching invariably associated with ESC-based therapies. The differentiation of human iPSCs into a hepatic lineage has recently been achieved (Sullivan et al. 2010); nonetheless limitations similar to ESC-derived differentiated cells (e.g., risk of teratoma formation) will remain. A series of rodent studies have shown that differentiated iPSCs can be used to rescue a diseased phenotype, namely sickle cell disease

(Hanna et al. 2007), Parkinson's disease (Wernig et al. 2008) and Haemophilia A (Xu et al. 2009), suggesting human iPSC-derived hepatocyte-like cells might be of therapeutic potential in metabolic liver disease. The realization of iPSC-derived liver stem cell therapy is some time away as yet; a more immediate use for this technology is likely to be *in vitro* cell-based assays invaluable in disease modelling and drug toxicology, particularly in studying the mechanisms of regeneration in liver cirrhosis.

Challenges of working with human tissue

> "We don't have to look for model organisms anymore because we are the model organism."
> Sydney Brenner, Nobel laureate for his work on *Caenorhabditis elegans*

The goal of virtually all biomedical research is to promulgate human applications and it is therefore logical to study human biology. However, the methods employed to identify and validate suitable therapeutic targets and potential new medicines have historically been almost entirely non-human, and the ability of such methods to predict efficacy and safety in humans has been unreliable. One does not have to look far to see the limitations of animal testing with tragedies such as the death of a gene transplant recipient (Hollon 2000) and the morbidity of the six volunteers testing TGN1412 (Dixit and Boelsterli 2007, Hansel et al. 2010, Cohen 2010) which was widely reported in the media. Thus, the use of human *in vitro* data to validate (or invalidate) animal studies will enable a more informed decision-making process as to their applicability and will greatly enhance the potential of successful translation into safe effective medicines. It is not the intention of the authors to suggest that human *in vitro* data alone will suffice, not least because isolated tissues cannot represent the intricate biological systems orchestrating *in vivo*, but the integration of human tissue research into drug discovery and therapeutics development is of paramount and direct relevance. Despite the undoubted value of human-based research, the acquisition of human tissues remains a significant challenge due largely to the many legal, ethical, logistical and practical issues that must be addressed in order to acquire such tissues.

Legal and ethical issues

It is beyond the scope of this chapter to discuss in detail the many facets of ethical issues in human tissue research and these will be dealt with in more detail in Chapter 11. Typically, adult human tissues may be acquired, with suitable consent, for research following medically-indicated surgery or organ donation post mortem. In the case of cadaveric donation, the use of tissue for transplantation always takes priority over research, and in surgical cases the requirement of tissue samples for pathology and diagnosis always takes precedence. Such prioritization, which is ethically correct, significantly reduces tissue availability for research. Furthermore, local research ethics committees often do not approve of taking extra tissue for research during surgery. Whilst there are challenges with the acquisition of adult tissue for research, the use of foetal abortions and human embryos evokes much more anxiety and controversy. This is not only because of religious sensitivities but also moral perspectives on elective termination of pregnancies and the need to respect the human embryo whether it originates from an established stem cell line, an aborted or miscarried embryo, an unused *in vitro* fertilized embryo or a cloned embryo, and each carries its own ethical issues. Central to the ethical debate revolving around human tissue research are three areas of consideration: the balancing of individual rights and social benefit; the researcher's conflict of interest; and the efficacy and integrity of the consent process which should adhere to the principles of the Nuremberg Code and the Helsinki Declaration. Ethical principles and legal regulations are closely intertwined as illustrated by the controversies that followed exposure of the "organ retention scandals" in 2001 leading to extensive law reforms in the UK legal systems in the form of the Human Tissue Act 2004 and the Human Tissue (Scotland) Act 2006. These legislations, strictly supervised by the Human Tissue Authority, firmly espouse the need for agreement to be reached before tissue and/or organs are removed, retained or used for either therapeutic or research purposes (Campbell et al. 2008). Whilst the legal system serves to protect, in particular, vulnerable individuals from participating in non-consensual research and to maintain public confidence by ensuring that human tissue is used safely and ethically, the licenses introduced have been criticised as costly, time-consuming and overly restrictive which could lead to endangerment of human-based research (Cronin et al. 2011). Researchers in the United States wishing to work with humans or human tissues face similar hurdles as the process of gaining ethical approval from the Institutional Review Board has been described as difficult and burdensome, and it is reported to be statistically less likely to get human-based research funded by the National Institutes of Health and other funding agencies than animal-based research (Infectious Diseases Society of America 2009, Rice 2011).

Logistical and infrastructural issues

The logistics of working with human tissue also provide challenges. Given the unpredictable nature of cadaveric availability and induced labour abortion, one cannot plan for exactly when human tissues may be available for processing and experimentation. For tissues that are to be prepared for storage, for example by freezing in liquid Nitrogen or by formalin fixation, medical intermediaries trained in the use of specific tissue-processing standard operating procedures can be recruited. However, for many other applications such as tissue/cell culture and genomics studies, fresh human tissue must be processed as quickly as possible to achieve satisfactory tissue/cell viability and RNA/DNA preservation. To ensure that high-quality human tissues and research services are delivered with maximum efficiency and minimum waste, an effective infrastructure must be in place, which in the UK takes the shape of research collaboration between the pharmaceutical and biomedical industries, universities and the NHS. This collaboration could potentially streamline the research contracting process (micra [Internet] 2011) and is particularly vital at a time of economic downturn when research is constrained by reductions in public funding. Systematic collection of tissue for research is not part of the core activity of the cash-limited NHS and a strategic framework taking into account the full economic costs has to be deliberated to accommodate such research activities. There are several areas of considerations: (i) training of clinical staff for their extended role, such as liaison/specialist nurses providing face-to-face consent interviews with donors and/or families, and histopathologists or biomedical scientists offering advice on where surplus fresh tissue is available for research without compromising the patient's clinical care; (ii) coordinated support from laboratory and mortuary staff; (iii) access to premises and staff availability out-of-hours; (iv) probity and integrity of the NHS public service; and (v) clinical, corporate and research governance in place to national and international standards (Womack and Gray 2007). The readers are referred to a first-hand account by Womack and Gray in their experience in establishing a public sector and commercial human tissue bank (Womack and Gray 2007).

The science

A key challenge for human tissue research lies in the expectations of the scientific community. The use of inbred species of rodents, transgenics, or recombinant humanized animals as experimental models allows consistent and reproducible data to be generated efficiently as researchers have better control over variables. Although these methods might appear elegant and produce results with tight margins of error, the amount of basic animal-based research being translated into human treatments appears to be at an all time low (Rice 2011). This is partly because of suboptimal animal models to represent or fully recapitulate human conditions, but also even in the presence of a reasonable model humans often respond differently to insults and interventions. As humans are outbred with vast genetic diversity and have infinitely variable external parameters that are impossible to control for, the level of standardization achievable in animal-based tests systems are unattainable when working with humans. This human variability demands well-designed *in vitro* experiments that are adequately powered with rigorous controls, flexibility within the research team, and availability of comprehensive clinical data to explain any anomalies or aberrant results.

References

Asahina K, K Teramoto and H Teraoka. 2006. Embryonic stem cells: hepatic differentiation and regenerative medicine for the treatment of liver disease. Curr Stem Cell Res Ther. 1(2): 139–56.

Aurich H, M Sgodda, P Kaltwasser et al. 2009. Hepatocyte differentiation of mesenchymal stem cells from human adipose tissue *in vitro* promotes hepatic integration *in vivo*. Gut. 58(4): 570–81.

Banas A, T Teratani, Y Yamamoto et al. 2007. Adipose tissue-derived mesenchymal stem cells as a source of human hepatocytes. Hepatology. 46(1): 219–28.

Barker N, JH van Es, J Kuipers et al. 2007. Identification of stem cells in small intestine and colon by marker gene Lgr5. Nature. 449(7165): 1003–1007.

Campbell AV, SAM McLean, K Gutridge et al. 2008. Human tissue legislation: listening to the professionals. J Med Ethics. 34(2): 104–108.

Chen Y, H Zhou, AL Sarver et al. 2010. Hepatic differentiation of liver-derived progenitor cells and their characterization by microRNA analysis. Liver Transpl. 16(9): 1086–1097.

Cohen AF.2010. Developing drug prototypes: pharmacology replaces safety and tolerability? Nat Rev Drug Discov. 9(11): 856–865.

Cronin AJ, ML Rose, JH Dark et al. 2011. British transplant research endangered by the Human Tissue Act. J Med Ethics. 37(8): 512–514.

Dalgetty DM, CN Medine, JP Iredale et al. 2009. Progress and future challenges in stem cell-derived liver technologies. Am. J. Physiol. Gastrointest. Liver Physiol. 297(2): G241–248.

Dixit R and UA Boelsterli. 2007. Healthy animals and animal models of human disease(s) in safety assessment of human pharmaceuticals, including therapeutic antibodies. Drug Discov. Today. (7-8): 336–342.

Elaut G, T Henkens, P Papeleu et al. 2006. Molecular mechanisms underlying the dedifferentiation process of isolated hepatocytes and their cultures. Curr Drug Metab. (6): 629–660.

Fausto N, JS Campbell and KJ Riehle. 2006. Liver regeneration. Hepatology. 43(2 Suppl 1): S45–53.

Fellous TG, S Islam, PJ Tadrous et al. 2009. Locating the stem cell niche and tracing hepatocyte lineages in human liver. Hepatology. 49(5): 1655–63.

Fisher RA and SC Strom. 2006. Human hepatocyte transplantation: worldwide results. Transplantation. 82(4): 441–449.

Fleig SV, SS Choi, L Yang et al. 2007. Hepatic accumulation of Hedgehog-reactive progenitors increases with severity of fatty liver damage in mice. Lab Invest. 87(12): 1227–39.

Forbes SJ and MR Alison. 2006. Side population (SP) cells: taking center stage in regeneration and liver cancer? Hepatology. 44(1): 23–6.

Gerlach JC, K Zeilinger and Ii JF Patzer. 2008. Bioartificial liver systems: why, what, whither? Regen Med. 3(4): 575–95.

Goodell MA, S McKinney-Freeman and FD Camargo. 2005. Isolation and characterization of side population cells. Methods Mol Biol. 290: 343–52.

Gouon-Evans V, L Boussemart, P Gadue et al. 2006. BMP-4 is required for hepatic specification of mouse embryonic stem cell-derived definitive endoderm. Nat Biotechnol. 24(11): 1402–11.

Hanna J, M Wernig, S Markoulaki et al. 2007. Treatment of sickle cell anemia mouse model with iPS cells generated from autologous skin. Science. 318(5858): 1920–1923.

Hansel TT, H Kropshofer, T Singer et al. 2010. The safety and side effects of monoclonal antibodies. Nat Rev Drug Discov. 9(4): 325–338.

Haridass D, Q Yuan, PD Becker et al. 2009. Repopulation efficiencies of adult hepatocytes, fetal liver progenitor cells, and embryonic stem cell-derived hepatic cells in albumin-promoter-enhancer urokinase-type plasminogen activator mice. Am. J. Pathol. 175(4): 1483–1492.

Hay DC, J Fletcher, C Payne et al. 2008. Highly efficient differentiation of hESCs to functional hepatic endoderm requires ActivinA and Wnt3a signaling. Proc Natl Acad Sci U S A. 105(34): 12301–6.

Heo J, VM Factor, T Uren et al. 2006. Hepatic precursors derived from murine embryonic stem cells contribute to regeneration of injured liver. Hepatology. 44(6): 1478–86.

Hollon T. 2000. Researchers and regulators reflect on first gene therapy death. Nat. Med. 6(1): 6.

Houlihan DD and PN Newsome. 2008. Critical review of clinical trials of bone marrow stem cells in liver disease. Gastroenterology. 135(2): 438–50.

Infectious Diseases Society of America. 2009. Grinding to a halt: the effects of the increasing regulatory burden on research and quality improvement efforts. Clin. Infect. Dis. 49(3): 328–335.

Inada M, D Benten, K Cheng et al. 2008. Stage-specific regulation of adhesion molecule expression segregates epithelial stem/progenitor cells in fetal and adult human livers. Hepatol Int. 2(1): 50–62.

Jung Y, KD Brown, RP Witek et al. 2008. Accumulation of hedgehog-responsive progenitors parallels alcoholic liver disease severity in mice and humans. Gastroenterology. 134(5): 1532–43.

Kisseleva T, H Uchinami, N Feirt et al. 2006. Bone marrow-derived fibrocytes participate in pathogenesis of liver fibrosis. J Hepatol. 45(3): 429–38.

Knight B, JE Tirnitz-Parker and JK Olynyk. 2008. C-kit inhibition by imatinib mesylate attenuates progenitor cell expansion and inhibits liver tumor formation in mice. Gastroenterology. 135(3): 969–79, 979 e1.

Kung JWC and SJ Forbes. 2009. Stem cells and liver repair. Curr Opin. Biotechnol. (5): 568–574.

Lagasse E, H Connors, M Al-Dhalimy et al. 2000. Purified hematopoietic stem cells can differentiate into hepatocytes *in vivo*. Nat Med. 6(11): 1229–34.

Levicar N, M Pai, NA Habib et al. 2008. Long-term clinical results of autologous infusion of mobilized adult bone marrow derived CD34+ cells in patients with chronic liver disease. Cell Prolif. 41 Suppl 1: 115–25.

Lyra AC. 2007. A pilot randomised controlled study used to evaluate efficacy of autologous bone marrow mononuclear cells transplantation in patients with advanced chronic liver disease (abstract). Hepatology. 46 (Suppl 1): 271A.

Mahieu-Caputo D, JE Allain, J Branger et al. 2004. Repopulation of athymic mouse liver by cryopreserved early human fetal hepatoblasts. Hum Gene Ther. 15(12): 1219–28.

McKenzie TJ, JB Lillegard and SL Nyberg. 2008. Artificial and bioartificial liver support. Semin Liver Dis. 28(2): 210–7.

micra [Internet]. [cited 2011 Aug 9];Available from: http: //www.nihr.ac.uk/infrastructure/Pages/micra.aspx

Newman RE, D Yoo, MA LeRoux et al. 2009. Treatment of inflammatory diseases with mesenchymal stem cells. Inflamm Allergy Drug Targets. 8(2): 110–123.

NHS Blood and Transplant [Internet]. [cited 2011 Jul 14]; Available from: http: //www.nhsbt.nhs.uk/news/2011/newsrelease040711.html

Nitou M, Y Sugiyama, K Ishikawa et al. 2002. Purification of fetal mouse hepatoblasts by magnetic beads coated with monoclonal anti-e-cadherin antibodies and their *in vitro* culture. Exp Cell Res. 279(2): 330–43.

Oertel M, A.Menthena, MD Dabeva and DA Shafritz. 2006. Cell competition leads to a high level of normal liver reconstitution by transplanted fetal liver stem/progenitor cells. Gastroenterology. 130(2): 507–20; quiz 590.

Oyagi S, M Hirose, M Kojima et al. 2006. Therapeutic effect of transplanting HGF-treated bone marrow mesenchymal cells into CCl4-injured rats. J Hepatol. 44(4): 742–8.

Plosch T, A Kosters, AK Groen et al. 2005. The ABC of hepatic and intestinal cholesterol transport. Handb Exp Pharmacol. (170): 465–82.

Puppi J and A Dhawan. 2009. Human hepatocyte transplantation overview. Methods Mol. Biol. 481: 1–16.

Rice MJ. 2011. The institutional review board is an impediment to human research: the result is more animal-based research. Philos Ethics Humanit Med. 6: 12.

Russo FP, MR Alison, BW Bigger et al. 2006. The bone marrow functionally contributes to liver fibrosis. Gastroenterology. 130(6): 1807–21.

Sahin MB, RE Schwartz, SM Buckley et al. 2008. Isolation and characterization of a novel population of progenitor cells from unmanipulated rat liver. Liver Transpl. 14(3): 333–345.

Sato T, JH van Es, HJ Snippert et al. 2011. Paneth cells constitute the niche for Lgr5 stem cells in intestinal crypts. Nature. 469(7330): 415–418.

Schmelzer E, E Wauthier and LM Reid. 2006. The phenotypes of pluripotent human hepatic progenitors. Stem Cells. 24(8): 1852–1858.

Schmelzer E, L Zhang, A Bruce, 2007. Human hepatic stem cells from fetal and postnatal donors. The Journal of Experimental Medicine. 204(8): 1973–87.

Sgodda M, H Aurich, S Kleist et al. 2007. Hepatocyte differentiation of mesenchymal stem cells from rat peritoneal adipose tissue *in vitro* and *in vivo*. Exp Cell Res. 313(13): 2875–86.

Smets F, M Najimi and EM Sokal. 2008. Cell transplantation in the treatment of liver diseases. Pediatr Transplant. 12(1): 6–13.

Spahr L, JF Lambert, L Rubbia-Brandt et al. 2008. Granulocyte-colony stimulating factor induces proliferation of hepatic progenitors in alcoholic steatohepatitis: a randomized trial. Hepatology. 48(1): 221–9.

Sullivan GJ, DC Hay, I-H Park et al. 2010. Generation of functional human hepatic endoderm from human induced pluripotent stem cells. Hepatology. 51(1): 329–335.

Sun C, DG Li, YW Chen et al. 2008. Transplantation of urokinase-type plasminogen activator gene-modified bone marrow-derived liver stem cells reduces liver fibrosis in rats. J Gene Med. 10(8): 855–66.

Takahashi K, K Tanabe, M Ohnuki et al. 2007. Induction of pluripotent stem cells from adult human fibroblasts by defined factors. Cell. 131(5): 861–72.

Takashimizu I, Y Tanaka, S Yoshie et al. 2009. Localization of Liv2 as an immature hepatocyte marker in EB outgrowth. Scientific World Journal 9: 190–9.

Tanimizu N, M Nishikawa, H. Saito et al. 2003. Isolation of hepatoblasts based on the expression of Dlk/Pref-1. J Cell Sci. 116(Pt 9): 1775–86.

Terrace JD, DC Hay, K Samuel et al. 2009. Side population cells in developing human liver are primarily haematopoietic progenitor cells. Exp Cell Res. 315(13): 2141–53.

Terrace JD, IS Currie, DC Hay et al. 2007. Progenitor cell characterization and location in the developing human liver. Stem cells and development. 16(5): 771–8.

van de Kerkhove MP, PP Poyck, T Deurholt et al. 2005. Liver support therapy: an overview of the AMC-bioartificial liver research. Dig Surg. 22(4): 254–64.

van der Meulen JHP, JD Lewsey, MF Dawwas et al. 2007. Adult orthotopic liver transplantation in the United Kingdom and Ireland between 1994 and 2005. Transplantation. 84(5): 572–579.

Watanabe T, K.Nakagawa, S Ohata et al. 2002. SEK1/MKK4-mediated SAPK/JNK signaling participates in embryonic hepatoblast proliferation via a pathway different from NF-kappaB-induced anti-apoptosis. Dev Biol. 250(2): 332–47.

Wernig M, JP Zhao, J Pruszak et al. 2008. Neurons derived from reprogrammed fibroblasts functionally integrate into the fetal brain and improve symptoms of rats with Parkinson's disease. Proc Natl Acad Sci USA. 105(15): 5856–61.

Womack C and NM Gray. 2007. Providing human tissue for research: 1996–2006. Pathobiology. 74(4): 212–217.

Wright N, L Samuelson, MH Walkup et al. 2008. Enrichment of a bipotent hepatic progenitor cell from naïve adult liver tissue. Biochem. Biophys. Res. Commun. 366(2): 367–372.

Xu D, Z Alipio, LM Fink et al. 2009. Phenotypic correction of murine hemophilia A using an iPS cell-based therapy. Proc Natl Acad Sci USA. 106(3): 808–13.

Yamanaka S. 2007. Strategies and new developments in the generation of patient-specific pluripotent stem cells. Cell Stem Cell. (1): 39–49.

Yannaki E, A Anagnostopoulos, D Kapetanos et al. 2006. Lasting amelioration in the clinical course of decompensated alcoholic cirrhosis with boost infusions of mobilized peripheral blood stem cells. Exp Hematol. 34(11): 1583–7.

Zajicek G, R Oren and M Weinreb. 1985. The streaming liver. Liver. 5(6): 293–300.

3

Role of Pluripotent Stem Cells in Regenerative Medicine

Eriona Hysolli,[1,a] *Xiao-Ling Zhou,*[1,b] *Renjing Liu,*[1,c] *Ji-Hyun Kim,*[1,d] *Brian Adams,*[1,e] *Gareth Sullivan,*[2] and *In-Hyun Park*[1,f,*]

Introduction

In the 1960s, James Till and Ernest McCulloch demonstrated that the injection of bone marrow cells into irradiated mice can give rise to multi-lineage hematopoietic colonies in the spleen (Becker et al. 1963, Till and McCulloch 1961). These "spleen colonies" were speculated to have arisen from a single bone marrow cell that can 1) self-renew and 2) differentiate into multiple lineages. Thus, the concept of a stem cell was born, and these two defining properties of stem cells still hold true today. Stem cells are classified into three broad categories depending on their differentiation potential: totipotent, pluripotent, and multipotent. The fertilization of the egg by a sperm produces a totipotent zygote, which further undergoes a series of cell division to become a blastocyst. Cells of the inner cell mass in

[1]Department of Genetics, Yale Stem Cell Center, Yale School of Medicine, 10 Amistad, New Haven, CT. 06520, USA.
[a]Email: eriona.hysolli@yale.edu
[b]Email: xiaoling.zhou@yale.edu
[c]Email: renjing.liu@yale.edu
[d]Email: ji-hyun.kim@yale.edu
[e]Email: brain.adams@yale.edu
[f]Email: inhyun.park@yale.edu
[2]MRC Centre for Regenerative Medicine, University of Edinburgh, 49 Little France Crescent, Edinburgh, EH16 4SB, UK; Email: garethsully@hotmail.com
*Corresponding author

the blastocyst are pluripotent and become an embryo proper composed of cells of all three germ layers. Multipotent stem cells are more restricted in their differentiation potential and are committed to differentiate into cells within a particular tissue type (Wagers and Weissman 2004).

Due to their self-renewing and differentiation potential, stem cells are considered as cellular resources for the treatment of several disorders that come as a result of disease or injury. As such, their role in regeneration is the subject of extensive research. In this chapter, we will review the potential of pluripotent stem cells for regenerative medicine (Fig. 3.1). We will first discuss the therapeutic potential of embryonic stem cells (ESCs) and provide an overview of methods for cellular reprogramming. Progress in the experimental and clinical applications of stem cells in regenerative medicine will be discussed. Challenges that need to be overcome to make stem cell-based therapy will be addressed.

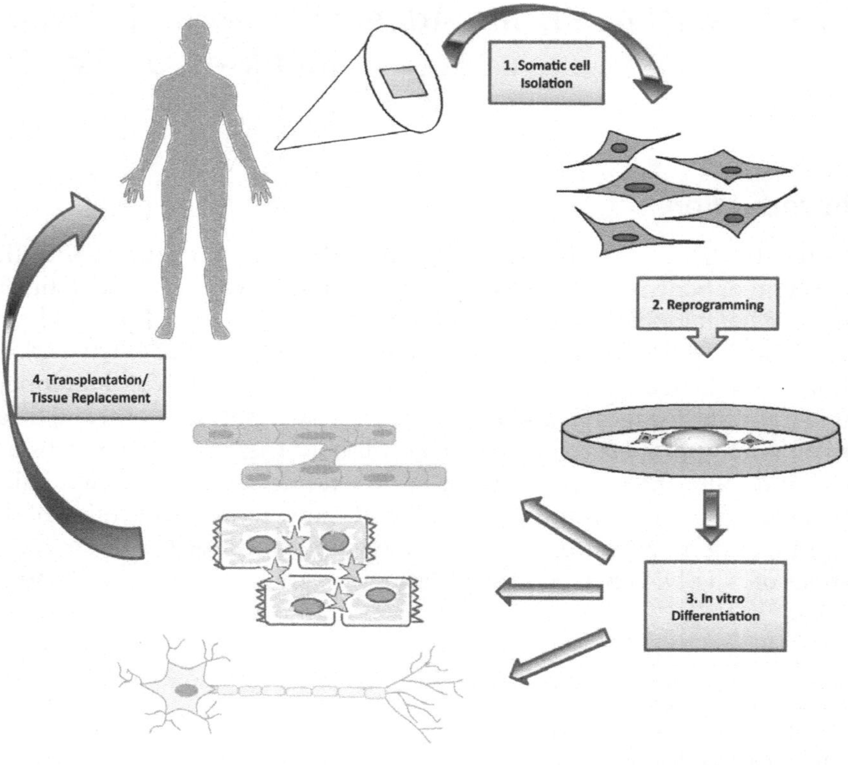

Figure 3.1. Overview of the step-wise application of pluripotent stem cells in regenerative medicine. Self-renewal and pluripotency render pluripotent ESCs or iPSCs as important cells for regenerative medicine. Somatic cells isolated from patients (Step 1) are to be reprogrammed via genenetic or non-genetic approach (Step 2). Reprogrammed iPSCs are in vitro differentiated into cells of interest (Step 3), which are used for the transplantation as autologous cells for the patients (Step 4).

Embryonic stem cells and their uses

ESCs are the *in vitro* derivatives of the inner cell mass of the blastocyst. ESCs have the attributes of self-renewal and pluripotency. The first ESC lines were established from mouse blastocysts in 1981 (Evans and Kaufman 1981, Martin 1981). The mouse ESCs (mESCs) grow on a feeder layer of non-proliferating mouse fibroblasts and maintain their undifferentiated state in the presence of leukemia inhibitory factor (LIF). mESCs give rise to germ-line chimaeras, and are able to do tetraploid complementation, two major criteria for testing the pluripotent capabilities of mESC lines (Bradley et al. 1984).

In 1998 the James Thomson group generated the first human ESCs (hESCs) from the inner cell mass of human blastocysts (Thomson et al. 1998). John Gearhart's group isolated human embryonic germ-cell lines at the same time (Shamblott et al. 1998). Human and mouse ESCs differ greatly in many major signaling pathways involved in maintaining their self-renewal and pluripotency properties (Ginis et al. 2004). The maintenance of self-renewal in mESCs involve the JAK-STAT signaling pathway, while that of hESCs occurs primarily through bFGF, IGF and MEK-ERK signaling (Darr and Benvenisty 2006).

ESCs offer an unprecedented opportunity to study an organism's early development, as well as promise for the field of regenerative medicine. The two characteristics of unlimited expansion and pluripotency make ESCs a very appealing resource for curing diseases (discussed in Section 4). Furthermore, ESCs can also be used to model human disease (Lensch and Daley 2006). As an example, human ESC lines were generated to model human diseases by disrupting ATM and p53 by targeting with bacterial artificial chromosomes (BACs) (Song et al. 2010).

Directing ESC differentiation in culture

ESCs can differentiate into cells of three germ layers: the ectoderm (skin and neural cells), the mesoderm (blood, bone and muscle), and the endoderm (visceral organs). These germ layers interact to form all the tissues and organs of the developing embryo. To date, many protocols have been established to differentiate ESCs to specific cell types of the three germ layers with the appropriate functional properties.

Differentiation of ESCs occurs spontaneously when cells are cultured as aggregates called embryoid bodies (EBs) in the absence of self-renewing factors, such as LIF (Leukemia inhibitory factor) for mESCs and bFGF for hESCs. Differentiation also ensues when ESCs are cultured directly on stromal cells, or when cultured as a monolayer over a gelatin or matrigel coat (Nishikawa et al. 1998). The addition of specific factors at distinct stages of ESC differentiation is important in directing cell-specific differentiation and enrichment. Growth factors such as Activin-A and Wnt3a mainly induce differentiation of ESCs down the endodermal lineage. Retinoic acid (RA), epidermal growth factor (EGF), bone morphogenic protein (BMP)-4, and basic fibroblast growth factor (bFGF) are known to elicit both ectodermal and mesodermal differentiation (Schuldiner et al. 2000). However, the effect of these factors depends on the species from which these ESCs are derived. For instance, BMPs and LIF together activate self-renewal of mouse ESCs, but hESCs undergo trophoblast or cardiac differentiation in the presence of BMPs (Xu et al. 2002).

Differentiation conditions for neuroectodermal lineage are well established. It has been observed that hESCs are prone to spontaneous formation of ectodermal derivatives on extended culture on the feeder cells (Reubinoff et al. 2001). All three major neural cell types (oligodendrocytes, astrocytes, and neurons) can be readily obtained by *in vitro* differentiation of ESCs. This has been confirmed by the use of gene expression analysis of specific neural markers. For example, oligodendrocyte precursor cells express the markers OLIG1, A2B5, SOX10, and NG2, while motor neuronal progenitors express BF1, HOXB4, NKX6-1/6-2, and OLIG1/2 (Erceg et al. 2009).

Mesodermal differentiation from ESCs is used to generate the hematopoietic, vascular, and cardiac lineages. Methods are well developed to produce both murine and human primitive hematopoietic cell lineages from ESCs (Orlovskaya et al. 2008). In the case of mESCs, the ectopic expression of HoxB4 promotes the engraftable blood formation from mESCs (Kyba M Fau 2002). However, the generation of robust levels of adult repopulating hematopoietic stem cells (HSCs) from genetically unmodified ESCs still requires further refinement (Huber 2010).

Differentiation of ESCs into cells of the endodermal lineage is the most challenging. This can be attributed to a number of factors. First, there are difficulties in discriminating between visceral and definitive endoderm since the markers of the definitive endoderm, such as Foxa2, Gata4, and Sox17, are also expressed in the visceral endoderm. Additionally, the identity and role of the factors needed for endodermal differentiation have not been fully explored (Keller 2005).

Advantages and shortcomings of using ESCs

Given their scalability and plasticity, ESCs represent a potential therapeutic solution to treat diseases that otherwise have no known cure. They also represent a potentially limitless source of cells to be used in regeneration and tissue repair via transplantation. But the current challenge is in obtaining suitable donor cells that are safe and effective. Tissue or cells derived from an allogeneic source of ESCs (i.e., from donated embryos) will most likely be rejected by the recipient through the activation of the immune system leading to serious problems, such as graft-versus-host disease. Customized ESCs from the patients themselves may be a solution to this. The generation of autologous, patient-specific ESCs through somatic cell nuclear transfer (SCNT) is one of the potential strategies. However, to date successful derivation of SCNT ESCs in humans has not been achieved. Thus, much effort has been used to find alternative strategies to derive patients-specific ESC-like pluripotent stem cells, which will be detailed in next section.

Chronology of reprogramming

The various cells found in our bodies arise from the successive differentiation steps following fertilization of the ovum. At each developmental stage, progenitor cells give rise to more mature cells, until the cells reach their terminally differentiated state. The epigenetic reversion of these differentiated cells into an embryonic state is so-called "reprogramming". The aim to isolate pluripotent stem cell-like cells for patients and to investigate the plasticity of cell fate has led to the study of reprogramming.

Somatic cell nuclear transfer

In 1952, Robert Briggs and Thomas J. King first performed an experiment where they injected the cells of late blastula into an enucleated egg of the frog *Rana pipiens* (Briggs and King 1952). The recipient eggs formed the complete blastulae, and occasionally more developed embryos. The ensuing series of nuclear transfer experiments involved using the terminally differentiated keratinizing cells from Xenopus laevis foot web explants (Gurdon et al. 1975). The egg nuclei were inactivated using UV light, and afterwards, the skin nuclei were injected into the inactivated eggs. Quite remarkably, the transplants underwent cleavage, and even tadpoles formed from "fertilizing" the egg with a diploid differentiated nucleus. These results demonstrated that oocytes contain the active materials that reprogram differentiated cells.

The first demonstration of successful reprogramming of mammalian cells was the generation of "Dolly" the sheep (Wilmut et al. 1997). Electrofusion of a mammary gland epithelium cell with an enucleated ewe egg followed by transplantation into a recipient animal was able to give rise to a fully developed live animal whose genetic composition was identical to the donor animal. The successful cloning of the first mouse named Cumulina was in 1998 by Teruhiko Wakayama and colleagues, thus providing the proof of the feasibility of the SCNT (Wakayama et al. 1998). Rigorous efforts succeeded in generating primate ESCs through SCNT, while the generation of human ESCs by SCNT is yet to be done (Byrne et al. 2007).

Like the egg cytoplasm that contains the factors necessary for reprogramming, ESCs themselves hold the potential to reprogram the differentiated cell. This is best exemplified in heterokaryon experiments where mESCs are fused with somatic human fibroblast, generating a hybrid cell with ESC-like characteristics (Cowan et al. 2005). The resultant cells are maintained in culture, and pass many of the pluripotency tests. However, these cells are still tetraploid with respect to chromosomal complement and thus challenge the clinical application.

Reprogramming with cellular extract

In 2005, Taranger and colleagues published the results of partial reprogramming of 293T and NIH3T3 cells by treatment with cell extract from a pluripotent embryonic carcinoma cell line NCCIT and a murine ESC line (Taranger et al. 2005). In brief, the human embryonic kidney cell line 293T and the NIH3T3 fibroblasts were first permeabilized, incubated with a pluripotent cell extract and resealed. The treated cells assumed a morphology closely resembling that of ESCs. The pluripotency gene OCT4/Oct4 was subsequently found to be upregulated and its promoter was demethylated. Furthermore, the partially reprogrammed 293T cells could be differentiated into other lineage cells in response to retinoic acid (RA) treatment. These results demonstrated that the cell extract of pluripotent cells contains the components that can promote reprogramming. Recently, the improvement of reprogramming method using cell extract from mESCs demonstrated the complete acquisition of pluripotency in mouse somatic cells (Cho et al. 2010).

A new era of reprogramming: induced pluripotent stem cells

In 2006, Kazutoshi Takahashi and Shinya Yamanaka set out to identify the genes important for reprogramming (Takahashi and Yamanaka 2006). By taking advantage of a unique mouse embryonic fibroblasts (MEFs) derived from a transgenic mouse expressing a neomycin resistance gene driven by an ESC specific gene namely, Fbx15. On transducing MEFs with the library of 24 candidate genes and culturing under conditions for maintaining pluripotency, colonies with an ESC-like morphology were selected with Neomycin. They further narrowed down the factors required for reprogramming to Oct4, Sox2, Klf4 and Myc. These ESC-like cells derived by the expression of just four transcription factors were named "induced pluripotent stem cells" (iPSCs).

Not very long after mouse iPSCs were generated, three independent groups described the reprogramming of human fibroblasts to pluripotency (Yu et al. 2007, Takahashi et al. 2007, Park et al. 2008). In one of these reports, the reprogramming was carried out by substituting KLF4 and c-MYC with two alternative pluripotency genes, namely LIN28 and NANOG, in addition to Oct4 and Sox2 (Yu et al. 2007). The generated human iPSCs were similar to hESCs with reference to expression of pluripotency genes and surface markers. They were positive for alkaline phosphatase and telomerase activity, and expressed the cell surface pluripotency markers SSEA-3, SSEA-4, Tra-1-60, and Tra-1-81. Human iPSCs readily generated teramtomas containing the three tissue lineages (ectoderm, endoderm and mesoderm). The kinetics of human iPSC generation is quite different from the mouse reprogramming process. Murine iPSC colonies appear within 1 wk, whereas human iPSC colonies generally appear no earlier than 3 wk.

Reprogramming by four factors is a process that can be applied to many cell types, but there are many drawbacks associated with this method. For instance, the reprogramming efficiency is relatively low (~0.1% or less). Thus, attempts were made to identify small molecules that improve the reprogramming efficiency. For example, the treatment of 5-aza-deoxycytidine (DNA methylation inhibitor), valproic acid and trichostatin A (histone deacetylase inhibitors), and Y27632 (ROCK inhibitor), as well as depletion of the expression of methyltransferases and tumor suppressors enhanced the reprogramming efficiency (Maherali and Hochedlinger 2008).

Reprogramming alternatives without transgene integration

Even though retroviral transgenes were shown to be silenced in iPSCs, chimeric mice from iPSCs developed tumors, this being attributed to reactivation of the c-Myc oncogene (Okita et al. 2007). c-Myc was found to be dispensable in the reprogramming process albeit at the expense of reduced reprogramming efficiency (Wernig et al. 2008). Moreover, even in the absence of c-Myc, random integration events of the provirus into the genome poses a safety issues in using iPSCs for potential activation of endogenous oncogenes. In order to bypass transgene integration issues concerted efforts to generate risk-free iPSCs have been a major focus.

An alternative to retroviral-based gene expression is the adenoviral delivery system that has a relatively low frequency of integration. Utilizing adenvirus to deliver the transgenes itself has associated issues, low reprogramming efficiency as compared to factors delivered by retrovirus. In fact, when Oct4 was delivered by an adenoviral vector, no iPS colonies were generated (Stadtfeld et al. 2008). Furthermore, a single polycistronic adenoviral vector containing the transgenes successfully reprograms somatic cells, but individually delivered transgenes do not (Okita et al. 2008). Thus, despite the advantage of no observable integration (mutagenesis) offered by the adenovirus system, the latter must be optimized in order to yield efficacious reprogramming.

Another method developed to induce reprogramming encompasses the generation of a cassette containing the Yamanaka factors flanked by loxP sites (Kaji et al. 2009). This virus-free system ensures one unique, albeit random, integration site, and subsequent excision of the trangenes upon expressing Cre recombinase. This system is applicable to derivation of virus free iPSCs. However, incomplete excision and spontaneous differentiation following removal of the factors are two main setbacks of this methodology. A similar method allowing excision of the factors after reprogramming, is achieved by utilizing an inducible *piggybac* vector (Woltjen et al. 2009). This transposon-based system allows the integration into multiple genomic sites, but transient expression of the transposase removes the trangenes after reprogramming. Naturally, simple serial transfections with the four factors has also been employed to generate ES-like cells, but with greatly reduced efficiency and random integration as the limitations of yet another virus-free method.

To eliminate non-endogenous genetic manipulations, protein and RNA based approaches have garnered interest for the future of iPSCs in stem cell therapy. In the protein-based approach, four reprogramming factors are expressed in either bacterial or mammalian systems, and then introduced into cells (Zhou et al. 2009). The main challenge is the ability to transport the protein cargo across the plasma membrane. To this end a number of basic motifs have been identified including poly-arginine or lysine motifs, and the HIV transactivator of transcription (TAT), a group collectively known as cell penetrating peptides. These have been utilized to facilitate the uptake of the macromolecules across the cell membrane. One major requirement of this system is the repeated application of these factors to the cells before the endogenous pluripotency program is activated. Recently RNA-based approaches have been developed using mRNA to initiate reprogramming. mRNA molecules of the four reprogramming factors were modified by introducing pseudouridylation, and a soluble interferon inhibitor to overcome innate antiviral response associated with unmodified mRNA (Warren et al. 2010). This approach is encouraging for the development of so-called "clean" iPSCs. The absence of genetic manipulation is appealing for potential therapeutic application (Stadtfeld and Hochedlinger 2010).

Potential use of stem cells for regenerative medicine

Self-renewal and pluripotency place pluripotent stem cells in a unique position in the field of regenerative medicine. To date, a number of studies have demonstrated the improvement of damaged organ through the stem cell transplantation (Aboody et al. 2011, Wu and Hochedlinger 2011, Laflamme and Murry 2011, Melton 2011). There are various degenerative diseases that would benefit from cell therapy and thus potentially restoring damaged or lost tissue. The following section describes a number of degenerative diseases and trials using stem cell transplantation to potentially improve disease state. We will describe the current status of these therapies in preclinical animal models and/or clinical applications and the challenges facing their translation into the clinic.

Myocardial regeneration

Myocardial infarction is an example of a degenerative diseases requiring cell transplantation. Cardiac remodeling such as thinning of the infarct wall and cardiac dilatation are results of myocardial infarction (Anversa et al. 1996, Jugdutt 2003). During the cardiac remodeling, abnormal cellular behavior is also observed, which includes apoptosis and necrosis of the cardiomyocytes, hypertrophy, fibrosis, and infiltration of inflammatory cells from the infracted heart (Jugdutt 2003, Kumar and Jugdutt 2003, Kumar et al. 2002). A number of cell sources, such as resident cardiac stem cells, bone marrow-derived stem cells, adipose-derived stem cells and skeletal myoblasts have been investigated as potential candidates for myocardial regenerative therapy (Krause et al. 2010, Rangappa et al. 2010, Singla 2010). The transplanted cells are not only the replacement of dead cells in the infracted heart but also the sources secreting factors to inhibit remodeling of the heart (Singla 2010). ESCs and iPSCs have emerged as one of the more promising sources of cardiac cells for transplantation. There have been many reported methods inducing the differentiation of pluripotent stem cells into cardiomyocyte or cardiac progenitor cells that not only share molecular markers with primary cardiomyocytes, but also exhibit similar electrophysiology and mechanical phenotypes (Freund and Mummery 2009). Currently, long-term therapeutic effect of donor cells for cardiac regeneration is a major issue for clinical application.

Kidney regeneration

The kidney is a mature organ that has a low-proliferative capacity and a very complex structure composed of several types of cells. These characteristics make the regeneration of the kidney extremely challenging (Iwatani and Imai 2010). Various types of stem cells have been examined for their therapeutic potential in kidney repair. These include bone marrow mesenchymal stem cells (BM-MSCs), which have been demonstrated to invoke positive effects with regard to repair of the damaged kidney (Krause and Cantley 2005, Morigi et al. 2004). Despite the low engraftment potential (3~20%), BM-MSC treated kidney recovered better than the non-treated group (Kale et al. 2003). This result suggests that BM-MSCs may provide a clement environment for the endogenous cells to recover by providing the necessary cytokines such as vascular endothelial growth factor (VEGF), hepatocyte growth factor (HGF), and insulin-like growth factor (IGF) (Togel et al. 2005). Likewise, MSC-like cells from both ESC and iPSC will potentially be a powerful resource for kidney regeneration. (Kim and Dressler 2005).

Stem cell therapy for diabetes

The characteristic of diabetes mellitus, hyperglycemia, is caused by either the deficiency of insulin or insulin resistance in body. In the case of insulin deficiency, this can be alleviated by an injection of insulin, but this is only temporally applicable and can potentially bring side effects such as lethal hypoglycemia. One solution is pancreas or islets cell transplantation as a final resort for diabetic patients and there has been remarkable success of insulin independency through organ transplantation. However, both number of donor tissue/organs and islets is a major limitation (Mishra et al. 2010).

Recently, insulin producing beta cells were generated from a number of stem cell types including embryonic, adult, hematopoietic, and mesenchymal stem cells. Transplantation of the *in vitro* derived beta cells is well considered for cell therapy for diabetes. However, proper modulation of the immune system is also important for the treatment of diabetes. Since MSCs have an immunomodulatory effect as well as the ability to differentiate into insulin-producing cells, MSCs are great candidate cells for treating diabetes (Vija et al. 2009). It has been demonstrated that beta-cells have been robustly differentiated from both ESCs and iPSCs, and thus will provide a potential novel avenue for the generation autologous β-cells for patients (Segev et al. 2004, Zhang et al. 2009).

Stem cell for liver regeneration

The liver has a comparatively high regenerative ability. Resident hepatic progenitor cells exist in liver, and the regenerative capacity is increased by a stimulus such as partial hepatectomy (Kuwahara et al. 2008). In addition, hepatic injury induces the mobilization of bone marrow-derived HSCs (Lemoli et al. 2006), which support the liver regeneration by enhancing vascular remodeling, macrophage-led matrix remodeling, and immune modulation (Stutchfield et al. 2010). However, severe liver injury requires more than what the resident stem cell can provide. The only cure for terminally ill patients of cirrhosis or hepatic carcinoma is an orthotopic liver transplant. Human hepatocyte transplantation is considered as an alternative to the liver transplantation. But, the limited supply of primary human hepatocytes has necessitated the pursuit of other sources. Transplantation of fetal liver precursor cells has been also studied for regeneration of the liver, but efficient liver regeneration was not observed (Haridass et al. 2009). MSCs can also differentiate into hepatocyte-like cells (Ikeda et al. 2008, Zheng et al. 2008). Mobilization and harvest of autologous bone marrow-derived stem cells by granulocyte colony stimulating factor (G-CSF) are actively studied as a promising cell candidate to prevent immune rejection and support damaged liver tissue (Stutchfield et al. 2010). However, MSCs contribute to the liver fibrosis, rendering MSCs not an ideal cell type for liver regeneration (Russo et al. 2006).

Hepatic-lineage cells are efficiently generated from both hESCs and iPSCs by combination of cytokines and grow factors, such as activin-A, oncostatin-M, hepatocyte growth factor (HGF), and fibroblast growth factors (FGFs) (Hay et al. 2008, Si-Tayeb et al. 2010, Sullivan et al. 2010). When they were transplanted into injured liver tissue, engraftment and differentiation into hepatocyte-like cells were observed, and showed some recovery of hepatic function in murine model (Agarwal et al. 2008, Moriya et al. 2008). The engraftment of hepatocytes derived from hESCs or iPSCs into larger animal models will be next step to pursue in order to test their safety and function before clinical trials.

Stem cell therapy for neuronal diseases

Patients with spinal cord injury (SCI) exhibit functional impairment such as sensory, motor and autonomic dysfunction, confusion of flow in fiber tracts and injury in peripheral circulation of spinal cord Schwab 2002). The primary damage in SCI includes the membrane and vascular disruption and hemorrhage. The secondary effect of SCI is much more severe and includes the axonal rim damage, induced active programmed cell death in spinal neurons and oligodendrocytes, and the glial scar formation (Sahni and Kessler 2010, Balentine 1988, Fawcett and Asher 1999, Chu et al. 2008). Transplantation of ESC-derived neuronal stem cells (NSCs) into the spinal cord has shown improvement to the injured tissue via the differentiation of the transplanted donor cells into mature functional neurons (Keirstead et al. 2005, Sharp et al. 2010).

Stem cell-based therapy in ischemic stroke is an actively evolving field (Banerjee et al. 2011). NSCs from either fetal or adult tissue have been shown to have the differentiation potential to generate both neurons and astrocytes, and enhance functional recovery when transplanted into the region of ischemic stroke (Chu et al. 2004, Chu et al. 2003, Kelly et al. 2004). The pluripotent embryonic carcinoma cell line NTera-2 (NT2) has been used as cells to derived donor neuronal cells. Treatment with retinoic acid and mitotic inhibitors initiated neuronal differentiation of NT2 cells, which showed the detectable improvement of brain function and behavior when applied in the murine model of stroke (Kleppner et al. 1995). Again both hESCs and iPSCs derived from the patients' own cells will provide a new alternative source for ischemic stroke cell therapy.

Parkinson's disease (PD) is exemplified by the loss of midbrain dopaminergic neurons. Clinical trials using mesencephalic tissue obtained from aborted embryos provided "proof of principle" that neuronal replacement is possible in PD patients. The grafted tissue was reported to significantly improve motor function in some patients (Lindvall et al. 1990). The variability in results, in addition to a shortage of donor tissues, together with incidences of graft-induced dyskinesis make it unlikely that neural transplantation using human fetal tissue will become a routine treatment for PD. Therefore, there is a need for an alternative source of dopamine-producing neurons that can lead to both long-lasting and significant improvements in motor ability and to counteract disease progression. ESCs and iPSCs are strong candidates to fulfill this role. Differentiation procedures for generation of dopamine neurons from ESCs have been established (Perrier et al. 2004) and researchers are actively refining the current methods of dopamine neuron production from pluripotent stem cells (Andersson et al. 2006, Castelo-Branco et al. 2006, Sacchetti et al. 2009, Vazin et al. 2008). Recently, Zeng and colleagues (2010) successfully generated dopamine-producing neurons from human iPSCs derived from both blood and skin (Swistowski et al. 2010). In addition, these cells on transplantation became functional and the rats showed improvement in their motor skills. This study again gives the potential cell therapies for Parkinson's disease.

Conclusion

The enormous potential of hESCs and iPSCs to treat human diseases has energized the scientific community and the lay public alike. These cells hold great promise in regenerative biology, but many key barriers will need to be overcome before these cells can be safely and effectively be put into clinical practice. One fundamental challenge is the reproducible isolation of pure cell populations for transplantation. Despite the rapidily evolving area of differentiation of both from hESCs and iPSCs to different cell types, it is still very difficult to isolate homogeneous populations of the specific cell type required. In addition, the incomplete differentiation of pluripotent stem cells may lead to unwanted teratoma formation in the donor (Wu and Hochedlinger 2011). The issue of survival and proper migration of donor cells in recipients raises another big question mark. Thus, the optimization and innovation in transplantation methods are the immediate research direction for fulfillment of the cellular regenerative medicine. Issues with low cell survival, poor migration and integration of stem cells upon transplantations will need to be addressed, as well as immunological complications surrounding the use of non-autologous stem cells. In this respect, iPSCs may be a better alternative and can be used to develop individual-specific stem cell lines with their own set of unique genetic material.

Although the clinical application of stem cell-based therapies for regeneration of injured and/or diseased tissues will face many hurdles, the combined efforts of laboratories will bring stem cell-based therapies closer to the bedside than ever before. In the coming years, we will undoubtedly increase our understanding of the biological properties of stem cells and their derivatives, leading to novel developments and technologies which patients will benefit from.

Acknowledgement

This work is supported by Charles Hood Foundation, and Yale Liver Center (NIDDK P30-34989).

References

Aboody K, A Capela, N Niazi et al. 2011. Translating stem cell studies to the clinic for CNS repair: current state of the art and the need for a Rosetta Stone. Neuron. 70: 597–613.

Agarwal S, KL Holton and R Lanza. 2008. Efficient differentiation of functional hepatocytes from human embryonic stem cells. Stem Cells. 26: 1117–1127.

Andersson E, U Tryggvason, Q Deng et al. 2006. Identification of intrinsic determinants of midbrain dopamine neurons. Cell. 124: 393–405.

Anversa P, J Kajstura and G Olivetti. 1996. Myocyte death in heart failure. Curr Opin Cardiol. 11: 245–251.

Balentine JD. 1988. Spinal cord trauma: in search of the meaning of granular axoplasm and vesicular myelin. J Neuropathol Exp Neurol. 47: 77–92.

Banerjee S, D Williamson, N Habib et al. 2011. Human stem cell therapy in ischaemic stroke: a review. Age Ageing. 40: 7–13.

Becker AJ, EA McCulloch and JE Till. 1963. Cytological demonstration of the clonal nature of spleen colonies derived from transplanted mouse marrow cells. Nature. 197: 452–454.

Bradley A, M Evans, MH Kaufman et al. 1984. Formation of germ-line chimaeras from embryo-derived teratocarcinoma cell lines. Nature. 309: 255–256.

Briggs R and TJ King. 1952. Transplantation of living nuclei from blastula cells into enucleated frogs' eggs. Proc Natl Acad Sci USA. 38: 455–463.

Byrne JA, DA Pedersen, LL Clepper et al. 2007. Producing primate embryonic stem cells by somatic cell nuclear transfer. Nature. 450: 497–502.

Castelo-Branco G, KM Sousa, V Bryja et al. 2006. Ventral midbrain glia express region-specific transcription factors and regulate dopaminergic neurogenesis through Wnt-5a secretion. Mol Cell Neurosci. 31: 251–262.

Cho HJ, CS Lee, YW Kwon et al. 2010. Induction of pluripotent stem cells from adult somatic cells by protein-based reprogramming without genetic manipulation. Blood. 116: 386–395.

Chu K, M Kim, KI Park et al. 2004. Human neural stem cells improve sensorimotor deficits in the adult rat brain with experimental focal ischemia. Brain Res. 1016: 145–153.

Chu K, M Kim, SW Jeong et al. 2003. Human neural stem cells can migrate, differentiate, and integrate after intravenous transplantation in adult rats with transient forebrain ischemia. Neurosci Lett. 343: 129–133.

Cowan CA, J Atienza, DA Melton et al. 2005. Nuclear reprogramming of somatic cells after fusion with human embryonic stem cells. Science. 309: 1369–1373.

Darr H and N Benvenisty. 2006. Human embryonic stem cells: the battle between self-renewal and differentiation. Regenerative Medicine. 1: 317–325.

Erceg S, M Ronaghi and M Stojkovic. 2009. Human embryonic stem cell differentiation toward regional specific neural precursors. Stem Cells. 27: 78–87.

Evans MJ and MH Kaufman. 1981. Establishment in culture of pluripotential cells from mouse embryos. Nature. 292: 154–156.

Fawcett JW and RA Asher. 1999. The glial scar and central nervous system repair. Brain Res Bull. 49: 377–391.

Freund C and CL Mummery. 2009. Prospects for pluripotent stem cell-derived cardiomyocytes in cardiac cell therapy and as disease models. J Cell Biochem. 107: 592–599.

Ginis I, Y Luo, T Miura et al. 2004. Differences between human and mouse embryonic stem cells. Developmental Biology. 269: 360–380.

Gurdon JB, RA Laskey and OR Reeves. 1975. The developmental capacity of nuclei transplanted from keratinized skin cells of adult frogs. Journal of Embryology and Experimental Morphology. 34: 93–112.

Haridass D, Q Yuan, PD Becker et al. 2009. Repopulation efficiencies of adult hepatocytes, fetal liver progenitor cells, and embryonic stem cell-derived hepatic cells in albumin-promoter-enhancer urokinase-type plasminogen activator mice. Am J Pathol. 175: 1483–1492.

Hay DC, J Fletcher, C Payne et al. 2008. Highly efficient differentiation of hESCs to functional hepatic endoderm requires ActivinA and Wnt3a signaling. Proc Natl Acad Sci USA. 105: 12301–12306.

Huber TL. 2010. Dissecting hematopoietic differentiation using the embryonic stem cell differentiation model. International Journal of Developmental Biology. 54: 991–1002.

Ikeda E, K Yagi, M Kojima et al. 2008. Multipotent cells from the human third molar: feasibility of cell-based therapy for liver disease. Differentiation. 76: 495–505.

Iwatani H and E Imai. 2010. Kidney repair using stem cells: myth or reality as a therapeutic option? J Nephrol. 23: 143–146.

Jugdutt BI 2003. Remodeling of the myocardium and potential targets in the collagen degradation and synthesis pathways. Curr Drug Targets Cardiovasc Haematol Disord. 3: 1–30.

Kaji K, K Norrby, A Paca et al. 2009. Virus-free induction of pluripotency and subsequent excision of reprogramming factors. Nature. 458: 771–775.

Kale S, A Karihaloo, PR Clark et al. 2003. Bone marrow stem cells contribute to repair of the ischemically injured renal tubule. J Clin Invest. 112: 42–49.

Keirstead HS, G Nistor, G Bernal and O Steward. 2005. Human embryonic stem cell-derived oligodendrocyte progenitor cell transplants remyelinate and restore locomotion after spinal cord injury. J Neurosci. 25: 4694–4705.

Keller GM. 2005. Embryonic stem cell differentiation: emergence of a new era in biology and medicine. Genes and Development. 19: 1129–1155.

Kelly S, TM Bliss, AK Shah et al. 2004. Transplanted human fetal neural stem cells survive, migrate, and differentiate in ischemic rat cerebral cortex. Proc Natl Acad Sci USA. 101: 11839–11844.

Kim D and GR Dressler. 2005. Nephrogenic factors promote differentiation of mouse embryonic stem cells into renal epithelia. J Am Soc Nephrol. 16: 3527–3534.

Kleppner SR, KA Robinson, JQ Trojanowski et al. 1995. Transplanted human neurons derived from a teratocarcinoma cell line (NTera-2) mature, integrate, and survive for over 1 year in the nude mouse brain. J Comp Neurol. 357: 618–632.

Krause D and LG Cantley. 2005. Bone marrow plasticity revisited: protection or differentiation in the kidney tubule? J Clin Invest. 115: 1705–1708.

Krause K, C Schneider, KH Kuck et al. 2010. Stem cell therapy in cardiovascular disorders. Cardiovasc Ther. 28: e101–110.

Kumar D and BI Jugdutt. 2003. Apoptosis and oxidants in the heart. J Lab Clin Med. 142: 288–297.

Kumar D, H Lou and PK Singal. 2002. Oxidative stress and apoptosis in heart dysfunction. Herz. 27: 662–668.

Kuwahara R, AV Kofman, CS Landis et al. 2008. The hepatic stem cell niche: identification by label-retaining cell assay. Hepatology. 47: 1994–2002.

Kyba M, Perlingeiro Rc, Daley, G.Q. 2002. HoxB4 confers definitive lymphoid-myeloid engraftment potential on embryonic stem cell and yolk sac hematopoietic progenitors. Cell. 109(1): 29–37.

Laflamme MA and CE Murry. 2011. Heart regeneration. Nature. 473: 326–335.

Lemoli RM, L Catani, S Talarico et al. 2006. Mobilization of bone marrow-derived hematopoietic and endothelial stem cells after orthotopic liver transplantation and liver resection. Stem Cells. 24: 2817–2825.

Lensch MW and GQ Daley. 2006. Scientific and clinical opportunities for modeling blood disorders with embryonic stem cells. Blood. 107: 2605–2612.

Lindvall O, P Brundin, H Widner et al. 1990. Grafts of fetal dopamine neurons survive and improve motor function in Parkinson's diseases. Science. 247: 574–577.

Maherali N and K Hochedlinger. 2008. Guidelines and techniques for the generation of induced pluripotent stem cells. Cell Stem Cell. 3: 595–605.

Martin GR. 1981. Isolation of a pluripotent cell line from early mouse embryos cultured in medium conditioned by teratocarcinoma stem cells. Proceedings of the National Academy of Sciences of the United States of America. 78: 7634–7638.

Melton DA. 2011. Using stem cells to study and possibly treat type 1 diabetes. Philos Trans R Soc Lond B Biol Sci. 366: 2307–2311.

Mishra PK, SR Singh, IG Joshua et al. 2010. Stem cells as a therapeutic target for diabetes. Front Biosci. 15: 461–477.

Morigi M, B Imberti, C Zoja et al. 2004. Mesenchymal stem cells are renotropic, helping to repair the kidney and improve function in acute renal failure. J Am Soc Nephrol. 15: 1794–1804.

Moriya K, M Yoshikawa,Y Ouji et al. 2008. Embryonic stem cells reduce liver fibrosis in CCl4-treated mice. Int J Exp Pathol. 89: 401–409.
Nishikawa SI, S Nishikawa, M Hirashima et al. 1998. Progressive lineage analysis by cell sorting and culture identifies FLK1+VE-cadherin+ cells at a diverging point of endothelial and hemopoietic lineages. Development. 125: 1747–1757.
Okita K, M Nakagawa, H Hyenjong et al. 2008. Generation of mouse induced pluripotent stem cells without viral vectors. Science. 322: 949–953.
Okita K, T Ichisaka and S Yamanaka. 2007. Generation of germline-competent induced pluripotent stem cells. Nature. 448: 313–317.
Orlovskaya I, I Schraufstatter, J Loring et al. 2008. Hematopoietic differentiation of embryonic stem cells. Methods. 45: 159–167.
Park IH, R Zhao, JA West et al. 2008. Reprogramming of human somatic cells to pluripotency with defined factors. Nature. 451: 141–146.
Perrier AL, V Tabar, T Barberi et al. 2004. Derivation of midbrain dopamine neurons from human embryonic stem cells. Proc Natl Acad Sci USA. 101: 12543–12548.
Rangappa S, R Makkar and J Forrester. 2010. Review article: current status of myocardial regeneration: new cell sources and new strategies. J Cardiovasc Pharmacol Ther. 15: 338–343.
Reubinoff BE, P Itsykson, T Turetsky et al. 2001. Neural progenitors from human embryonic stem cells. Nature Biotechnology. 19: 1134–1140.
Russo FP, MR Alison, BW Bigger et al. 2006. The bone marrow functionally contributes to liver fibrosis. Gastroenterology. 130: 1807–1821.
Sacchetti P, KM Sousa, AC Hall et al. 2009. Liver X receptors and oxysterols promote ventral midbrain neurogenesis *in vivo* and in human embryonic stem cells. Cell Stem Cell. 5: 409–419.
Sahni V and JA Kessler. 2010. Stem cell therapies for spinal cord injury. Nature Review Neurology. 6: 363–372.
Schuldiner M, O Yanuka, J Itskovitz-Eldor et al. 2000. Effects of eight growth factors on the differentiation of cells derived from human embryonic stem cells. Proceedings of the National Academy of Sciences of the United States of America. 97: 11307–11312.
Schwab ME. 2002. Repairing the injured spinal cord. Science. 295: 1029–1031.
Segev H, B Fishman, A Ziskind and et al. 2004. Differentiation of human embryonic stem cells into insulin-producing clusters. Stem Cells. 22: 265–274.
Shamblott MJ, J Axelman, S Wang et al. 1998. Derivation of pluripotent stem cells from cultured human primordial germ cells. Proceedings of the National Academy of Sciences of the United States of America. 95: 13726–13731.
Sharp J, J Frame, M Siegenthaler et al. 2010. Human embryonic stem cell-derived oligodendrocyte progenitor cell transplants improve recovery after cervical spinal cord injury. Stem Cells. 28: 152–163.
Singla DK 2010. Stem cells in the infarcted heart. J Cardiovasc Transl Res. 3: 73–78.
Si-Tayeb K, FK Noto, M Nagaoka et al. 2010. Highly efficient generation of human hepatocyte-like cells from induced pluripotent stem cells. Hepatology. 51: 297–305.
Song H, SK Chung and Y Xu. 2010. Modeling disease in human ESCs using an efficient BAC-based homologous recombination system. Cell Stem Cell. 6: 80–89.
Stadtfeld M and K Hochedlinger. 2010. Induced pluripotency: history, mechanisms, and applications. Genes Dev. 24: 2239–2263.
Stadtfeld M, M Nagaya, J Utikal et al. 2008. Induced pluripotent stem cells generated without viral integration. Science. 322: 945–949.
Stutchfield BM, SJ Forbes and SJ Wigmore. 2010. Prospects for stem cell transplantation in the treatment of hepatic disease. Liver Transpl. 16: 827–836.
Sullivan GJ, DC Hay, I-H Park et al. 2010. Generation of functional human hepatic endoderm from human induced pluripotent stem cells. Hepatology. 51: 329–335.

Swistowski A, J Peng, Q Liu et al. 2010. Efficient generation of functional dopaminergic neurons from human induced pluripotent stem cells under defined conditions. Stem Cells. 28: 1893–1904.

Takahashi K and S Yamanaka. 2006. Induction of pluripotent stem cells from mouse embryonic and adult fibroblast cultures by defined factors. Cell. 126: 663–676.

Takahashi K, K Tanabe, M Ohnuki et al. 2007. Induction of pluripotent stem cells from adult human fibroblasts by defined factors. Cell 131: 861–872.

Taranger CK, A Noer, AL Sorensen et al. 2005. Induction of dedifferentiation, genomewide transcriptional programming, and epigenetic reprogramming by extracts of carcinoma and embryonic stem cells. Mol Biol Cell. 16: 5719–5735.

Thomson JA, J Itskovitz-Eldor, SS Shapiro et al. 1998. Embryonic stem cell lines derived from human blastocysts. Science. 282: 1145–1147.

Till JE and EA McCulloch. 1961. A direct measurement of the radiation sensitivity of normal mouse bone marrow cells. Radiation Research. 14: 213–222.er.

Togel F, Z Hu, K Weiss et al. 2005. Administered mesenchymal stem cells protect against ischemic acute renal failure through differentiation-independent mechanisms. Am J Physiol Renal Physiol. 289: F31–42.

Vazin T, J Chen, CT Lee et al. 2008. Assessment of stromal-derived inducing activity in the generation of dopaminergic neurons from human embryonic stem cells. Stem Cells. 26: 1517–1525.

Vija L, D Farge, JF Gautier et al. 2009. Mesenchymal stem cells: Stem cell therapy perspectives for type 1 diabetes. Diabetes Metab. 35: 85–93.

Wagers AJ and IL Weissman. 2004. Plasticity of adult stem cells. Cell. 116: 639–648.

Wakayama T, ACF Perry, M Zuccotti et al. 1998. Full-term development of mice from enucleated oocytes injected with cumulus cell nuclei. Nature. 394: 369–374.

Warren L, PD Manos, T Ahfeldt et al. 2010. Highly efficient reprogramming to pluripotency and directed differentiation of human cells with synthetic modified mRNA. Cell Stem Cell. 7: 618–630.

Wernig M, A Meissner, JP Cassady et al. 2008. c-Myc is dispensable for direct reprogramming of mouse fibroblasts. Cell Stem Cell. 2: 10–12.

Wilmut I, AE Schnieke, J McWhir et al. 1997. Viable offspring derived from fetal and adult mammalian cells. Nature. 385: 810–813.

Woltjen K, IP Michael, P Mohseni et al. 2009. PiggyBac transposition reprograms fibroblasts to induced pluripotent stem cells. Nature. 458: 766–770.

Wu SM and K Hochedlinger. 2011. Harnessing the potential of induced pluripotent stem cells for regenerative medicine. Nat Cell Biol. 13: 497–505.

Xu RH, X Chen, DS Li, et al .2002. BMP4 initiates human embryonic stem cell differentiation to trophoblast. Nature Biotechnology. 20: 1261–1264.

Yu J, MA Vodyanik, K Smuga-Otto et al. 2007. Induced pluripotent stem cell lines derived from human somatic cells. Science. 318: 1917–1920.

Zhang D, W Jiang, M Liu et al. 2009. Highly efficient differentiation of human ES cells and iPS cells into mature pancreatic insulin-producing cells. Cell Res. 19: 429–438.

Zheng YB, ZL Gao, C Xie et al. 2008. Characterization and hepatogenic differentiation of mesenchymal stem cells from human amniotic fluid and human bone marrow: a comparative study. Cell Biol Int. 32: 1439–1448.

Zhou H, S Wu, JY Joo et al. 2009. Generation of induced pluripotent stem cells using recombinant proteins. Cell Stem Cell. 4: 381–384.

4

Human Liver Development as a Template to Generate High Fidelity Models

Claire N. Medine,[1,a] *Janet W.C. Kung,*[1]
Catherine M. Payne,[1,b] *James R. Black,*[1,c]
Richard A. Anderson,[1,d] *James A. Ross,*[1,e]
John P. Iredale[1] and *David C. Hay*[1,f,*]

Introduction

Cellular niches are specialized microenvironments that integrate numerous stimuli. The complex interpretation of these signals is crucial for normal human development and tissue homeostasis in the neonate. In this chapter we will focus on human liver biology and the role played by Wnt signaling in the foetal and adult niche.

The Wnt signaling pathway is a complex network which consists of secreted signalling molecules that bind to specific receptors to activate downstream signalling cascades (Logan and Nusse 2004, Moon et al. 1997). The canonical Wnt pathway describes a series of events that occur when Wnt

[1]MRC Centre for Regenerative Medicine, University of Edinburgh, 49 Little France Crescent, Edinburgh, EH16 4SB, U.K.
[a]Email: cmedine@staffmail.ed.ac.uk
[b]Email: cateduff@btinternet.com
[c]Email: jblack2@staffmail.ed.ac.uk
[d]Email: r.anderson@hrsu.mrc.ac.uk
[e]Email: j.a.ross@ed.ac.uk
[f]Email: davehay@talktalk.net
*Corresponding author

proteins are presented to the surface of target cells and signal through the cell surface receptors, Frizzled (Fz)/low density lipoprotein (LDL) receptor related protein (LRP). Ling/FZ/LRP engagement leads to the activation of Dishevelled (Dsh) in the cytoplasm and recruitment of Axin to the plasma membrane. This inhibits the assembly of the destruction complex (Axin, Glycogen synthase kinase-3β (GSK-3), Adenomatous Polyposis Coli (APC), and Casein kinase (CKI)) promoting β-Catenin stability and subsequent nuclear translocation (Logan and Nusse 2004, Moon et al. 1997). Nuclear β-Catenin interacts with transcription factors such as the T cell-specific factor/lymphoid enhancer-binding factor (TCF/LEF) where it regulates gene transcription (Arce et al. 2006, Clevers 2006, Wu and Nusse 2002, Kimelman and Xu 2006). It is important to acknowledge that non-canonical Wnt pathways also exist. Signal transduction in this case is independent of β-catenin, and instead elicit an increase in intracellular calcium via interactions with protein kinase C or activate small Rho GTPases, including c-Jun N terminus kinase (Kuhl et al. 2000, Nateri et al. 2005, Topol et al. 2003, Westfall et al. 2003). For reasons of simplicity, we will focus solely on the canonical Wnt signaling pathway in this chapter.

Our interest lies in human liver development. During human development three germ layers are formed (endoderm, mesoderm and ectoderm) that contribute to all the cell types found in the embryo proper. The liver is formed from the endoderm which is produced during gastrulation. At this point undifferentiated epiblast cells migrate through the primitive streak, forming mesoderm and endoderm (Lawson et al. 1991, Kinder et al. 2001, Thomson et al. 1998). A key determinant in cell fate choice between endoderm and mesoderm is Nodal signalling (Schier 2003) and enhances the expression of a core group of endodermal associated transcription factors; including HNF3, Gata 4 and Gata 6 (Bossart and Zaret 1998, Zaret 1999). These factors are responsible for chromatin decompaction (Cirillo et al. 2002, Zhao and Duncan 2005, Zhao et al. 2005) which renders the foregut endoderm competent to respond to stimuli from the surrounding mesenchyme, culminating in liver bud formation (Zhao and Duncan 2005, Jung et al. 1999, Rossi et al. 2001, Shin et al. 2007). Following establishment of the liver bud, the resident bipotential stem cell population (the hepatoblast) differentiate to hepatic parenchyma or biliary duct epithelia under guidance of an activin signalling gradient regulated by the transcription factors HNF-6 and OC-2 (Clotman et al. 2002, Clotman et al. 2005).

During embryonic development and in the adult, the correct temporal and spatial expression of Wnt signaling factors also play essential roles in a number of processes; from primitive streak (PS) formation (Lawson et al. 1991, Kinder et al. 2001, Thomson et al. 1998, Gadue et al. 2006, Hay et al. 2008) and liver bud formation (Ober et al. 2006) to organ homeostasis in the adult (Tan et al. 2006). Therefore, the correct regulation of Wnt signalling is key to cell physiology and there are a number of reports which demonstrate that dysregulation of this pathway is involved in a variety of human malignancies (Ding et al. 2009, Hu et al. 2009, Wei et al. 2009, Tang et al. 2009, Rhee et al. 2002, Mazieres et al. 2004, Weeraratna et al. 2002). Moreover, there is increasing evidence that supports the notion that Wnt family members can act as both tumour suppressors and activators in the liver (Ding et al. 2009, Hu et al. 2009, Wei et al. 2009, Tang et al. 2009, Shih et al 2006, Ueno et al. 2008, Urakami et al. 2006a, Urakami et al. 2006b, Urakami et al. 2006c, Deng et al. 2010, Koch et al. 2005, Wirths et al. 2003, Lustig et al. 2002).

The purpose of this chapter is to highlight the instructive role that human liver tissue has played in developing a better understanding of Wnt signaling. During liver development the ability to harness this information and create valuable *in vitro* models from stem cells may permit the development of novel medicines and/or regenerative strategies to treat human liver disease.

Wnt signaling and hepatocyte differentiation

During gastrulation endoderm and mesoderm are formed from a common precursor, mesendoderm. Yamaguchi et al. demonstrated that Wnt3 and Wnt3a are expressed in the PS of mice and are necessary for mesoderm specification (Yamaguchi et al. 1999). This work was complemented by Haegel et al. (1995) who demonstrated that although β-Catenin null embryos were able to form implantable blastocysts that developed into egg cylinder stage embryos, they were unable to form the primitive streak, and therefore mesoderm or endoderm (Haegel et al. 1995) (Liu et al. 1999).

During development, the specification of the anterior region of the PS to mesoderm or endoderm depends on the duration and magnitude of Nodal signalling. To investigate the role of Wnt3 and Activin signalling in primitive streak, mesoderm and endoderm formation in an embryonic stem (ES) cell model, Gadue et al. established a murine ES cell line that expressed CD4 from the *foxa2* locus in addition to GFP from the brachury locus. Using this model, they also showed that simultaneous Wnt3 and Activin were essential for ES cell PS induction, and sustained Activin signalling was found to stimulate endoderm and early hepatic commitment (Gadue et al. 2006). More recently, we demonstrated that Wnt3a and Activin A signalling was important in definitive endoderm and hepatic differentiation from human embryonic stem cell (hESC) and induced pluripotent stem cells (iPSCs) (Hay et al. 2008, Sullivan et al. 2010). The process of hepatocyte differentiation was most efficient in hESC and IPSC cultures which did not possess feeders or stromal cell mesenchyme (SCM) (Sullivan et al. 2010, Fletcher et al. 2008). Stem cells maintained in a co-culture environment, with mesenchyme, did not respond well to differentiation cues and maintained their colony structure and characteristic cell size and shape during the process indicating a block in cellular differentiation. Moreover, stem cell transcripts Oct-4 and Nanog could be detected by PCR following the differentiation procedure (Fletcher et al. 2008). These experiments were an important demonstration of the changes in cell biology that arose in a simple co-culture niche. It was therefore important to understand how we could reinstate efficient hepatic endoderm differentiation in our model.

We discovered that the negative effect of SCM on hESC differentiation could be reversed following Wnt3a pre-treatment of hESC/SCM co-cultures prior to initation of cell differentiation (Fletcher et al. 2008). To better understand the mechanism behind our model a Wnt signaling array was employed to measure the expression of major Wnt signaling components. hESC RNA was isolated from both co-culture and pure hESC cultures. The most significant gene expression changes within the hESC populations during co-culture, were the up regulation of WIF-1 and dramatic reduction of Wnt3 and Wnt3a gene expression (Fig. 4.1A). The SCM was removed from co-cultures before analysis and did not express these factors, indicating that the SCM/hESC interaction likely resulted in increased levels of WIF-1 expression in hESC populations. Given that we are interested in stabilizing cell phenotype within our model and that mesenchymal cell populations have been shown to support long-term hepatic function, we studied this relationship further. hESC/SCM cultures were differentiated to hepatocyte like cells (HLCs) in the presence or absence of Wnt3a (Fletcher et al. 2008). As expected, the presence of Wnt3a pre-treatment hESC/SCM cultures exhibited efficient differentiation to HLCs with ~65% (± 4.9%) of the cells displaying albumin production (Fig. 4.1B). This was in contrast to cells which did not receive Wnt3a pre-treatment with ~0.6% (± 0.3%) expressing albumin (Fig. 4.1B). In light of the array data we wished to determine if attenuated HLC differentiation could be attributed to elevated levels of WIF-1 production. Therefore, hESC cultures were treated with WIF-1 prior to onset of Wnt3a driven hepatocyte differentiation. WIF-1 pre-treatment had a negative effect on hepatic differentiation with ~5 fold decrease in albumin positive HLCs [~13% (± 3.2%)] (Fig. 4.1B). In line with decreased HLC differentiation, we also observed a decrease in liver specific function with a reduction in fibronectin and fibrinogen secretion (Fig. 4.1C).

Wnt signaling during human liver development

Secreted Wnt proteins play an essential role in cellular differentiation. In the last decade, genetic analysis and multiple studies using embryonic tissue explants have provided valuable insight into the role of the Wnt/β-catenin signaling pathway in liver development [for a review see Thompson and

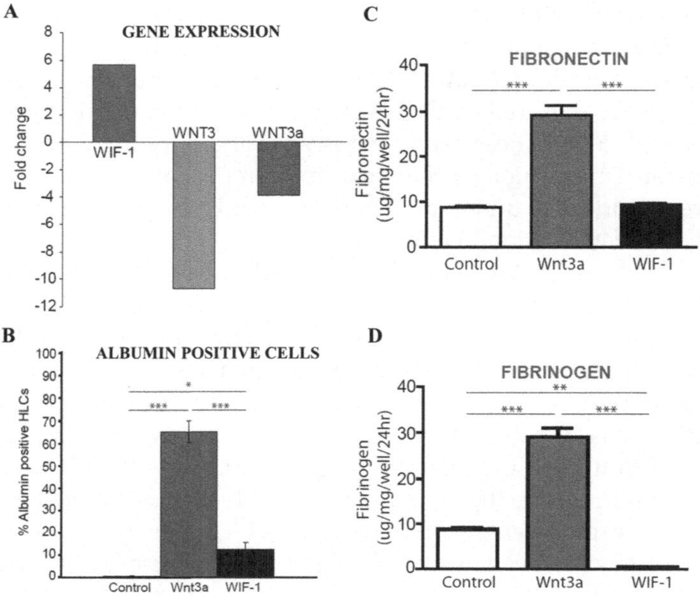

Figure 4.1. WIF-1 is an important factor in pluripotent stem cell differentiation potential.

A. The effect of SCM on expression of key components of the Wnt signalling pathway in hESC populations was measured using a Wnt super array (SA Biosciences). RNA from early passage hESCs maintained in stromal hESC cultures was compared to SCM free hESC populations and quoted as fold change over SCM free hESC populations. Wnt3 and Wnt3a gene expression were down regulated in the presence of stroma, while the Wnt antagonist WIF-1 showed a 6 fold increase.
B. hESC differentiation to hepatic endoderm proceeded normally in hESC and SCM co-cultures pre-treated with Wnt3a prior to differentiation. In the presence of SCM and WIF-1 prior to Wnt3a treatment, differentiation to hepatic endoderm was drastically reduced. Hepatic differentiation efficiency was measured by albumin expressing cell counts (n= 4 fields of view, ~200-1500 cells) and the mean percentage of hepatocyte like cells (HLCs) ± sem plotted. hESC differentiation to hepatic HLCs was drastically reduced in the presence of WIF-1 and SCM (~13% HLCs) when compared to cells cultured in Wnt3a prior to differentiation (~65% HLCs). Student t-tests were used to calculate the significance. The asterisks define the level of significance where *—significant, **—very significant and ***—extremely significant (p<0.0001). An IgG antibody (1:500) was used as a negative control. Immunostaining was recorded using an inverted Zeiss fluorescent microscope. hESC cells were differentiated as described previously (Fletcher et al. 2008).
C. 17 d post differentiation cell culture supernatants were collected and analyzed for serum protein production by Enzyme-Linked Immunosorbent Assay (ELISA). In line with cellular commitment, hepatic function was significantly reduced in the presence of SCM and WIF-1 pre-treatment. Wnt3a pre-treated cells, exhibited significantly higher level production of hepatic serum proteins Fibronectin and Fibrinogen than the control and WIF-1 pre-treated cells. Student t-tests were used to calculate the significance, where asterisks define the level of significance where *—significant, **—very significant and ***—extremely significant (p<0.0001). Data are expressed as the mean ± sem, n = 3 per treatment. Units are quoted as ug/mg/ per well over 24 hr. Supernatants were collected for export hepatic protein assays and measured as described in (Fletcher et al. 2008).

Monga 2007]. Monga and colleagues demonstrated a role for Wnt/β-catenin in early liver development by utilizing *in vitro* embryonic liver cultures (Monga et al. 2003). Additional studies by Micsenyi and colleagues (2004) used livers from numerous stages of gestation and in neonate to study β-Catenin expression, protein-protein interactions, localization, and regulation. During early liver development, peak β-Catenin expression was observed at E10–E12, followed by a decrease and reduction in β-Catenin expression and protein stability after E16 (Micsenyi et al. 2004), highlighting stage-specific regulation of β-Catenin expression and stability during liver development. Canonical Wnt signaling has also been shown to be essential for normal liver bud development in zebra fish (Ober et al. 2006). Studies revealed that a Wnt2b homologue expressed in the lateral plate mesoderm, directly adjacent to the liver forming endoderm, was important for liver bud formation. The essential role of Wnt signaling in adult liver biology has also been studied using β-Catenin conditional knockout murine models (Tan et al. 2006). The loss of β-Catenin expression led to a significant decrease in liver weight/body weight ratio, decrease in basal hepatocyte proliferation and increased apoptosis during liver regeneration (Tan et al. 2006). We have also examined Wnt3a expression during human liver development, specifically in the first (T1) and second (T2) trimester (Hay et al. 2008). Wnt3a expression was adjacent to developing portal structures in T1 liver, but was not detected in the parenchyma. However, in T2 livers Wnt3a expression was both portal and parenchymal, suggesting that Wnt3a parenchymal expression correlated with and likely influenced hepatocyte specification within the developing human liver (Hay et al. 2008).

Regulation of Wnt signaling

Wnt signaling is controlled by various antagonists that can be divided into two categories (Kawano and Kypta 2003). The first group includes Wnt inhibitory factor (WIF-1), the secreted frizzled-related protein (sFRP) family, and Cerberus which bind directly to Wnt ligands blocking their interaction with frizzled proteins. The second group includes members of the Dickkopf (Dkk) family which block Wnt signaling by inhibiting the formation of the Wnt-induced Frizzled–LRP5/6 complex (van Es et al. 2003, Semenov et al. 2001).

In order to further understand the regulation of Wnt signalling in human development and test the fidelity of our *in vitro* model, we examined WIF 1 expression in T1 and T2 liver. WIF-1 expression was detected at high levels in human T1 liver adjacent to developing hepatoblasts (Fig. 4.2A), indicating that the regulation of Wnt signaling was important in early liver development. In contrast, WIF-1 expression was not detected in T2 liver (Fig. 4.2B) nor the controls (Fig. 4.2C and D), suggesting that Wnt signaling was elevated in T2 liver as hepatoblasts committed to hepatocytes, in line with

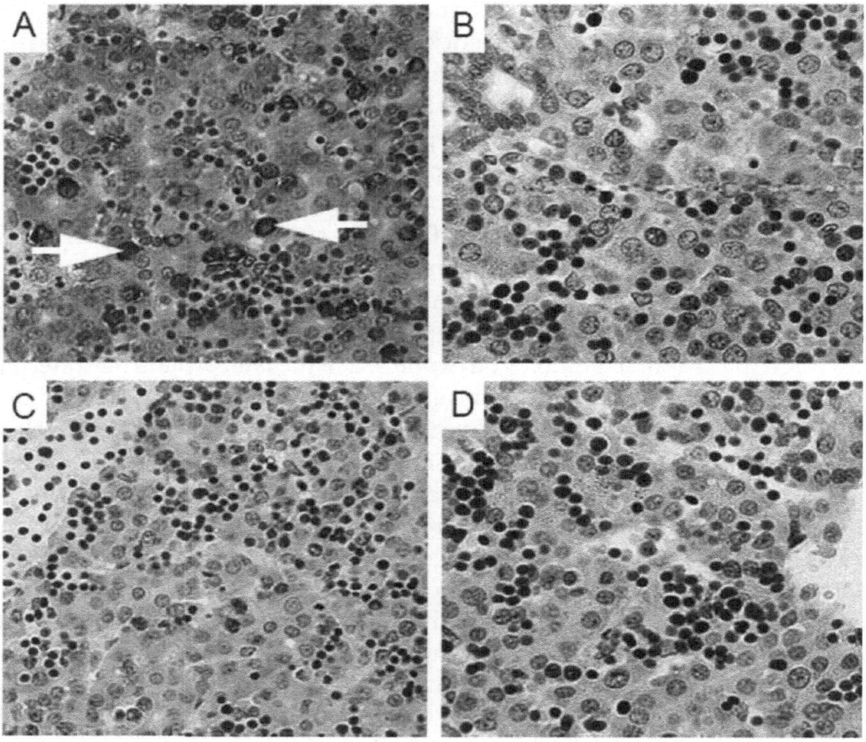

Figure 4.2. WIF-1 is an important factor in human liver development.
Wnt Inhibitory Factor-1 (WIF-1) immunostaining was carried out in human T1 (A) and T2 (B) embryonic liver paraffin embedded sections. WIF-1 expression was pronounced in areas adjacent to developing hepatoblasts in T1 liver (see white arrows, A). In contrast, WIF-1 expression was not detected in T2 liver. Isotype controls stained negative in T1 (C) and T2 (D) liver demonstrating specificity. Immunohistochemistry images were captured using a Leica FW4000 brightfield microscope. Images are at x20 magnification.

Color image of this figure appears in the color plate section at the end of the book.

our previous observations (Hay et al. 2008). We extended these studies to adult liver and similar to T2 liver we did not detect WIF-1 staining in the parenchyma or in control slides (Fig. 4.3A and B).

Given that the secreted antagonists of the Wnt signaling pathway, have been found to be dysregulated to a variety of cancers (Mazieres et al. 2004, Koch et al. 2005, Wirths et al. 2003, Lustig et al. 2002, Wissmann et al. 2003, Zhou et al. 1998, Suzuki et al. 2004, Lee et al. 2004, Shou et al. 2002) we were also interested in the relationship between WIF-1 expression and human liver cancer. We do not detect WIF-1 staining in normal adult liver (Fig. 4.3 A) nor controls (Fig. 4.3 B) but detected pronounced WIF-1 expression in tumours in the presence of a stromal component (Fig. 4.3C) but not in

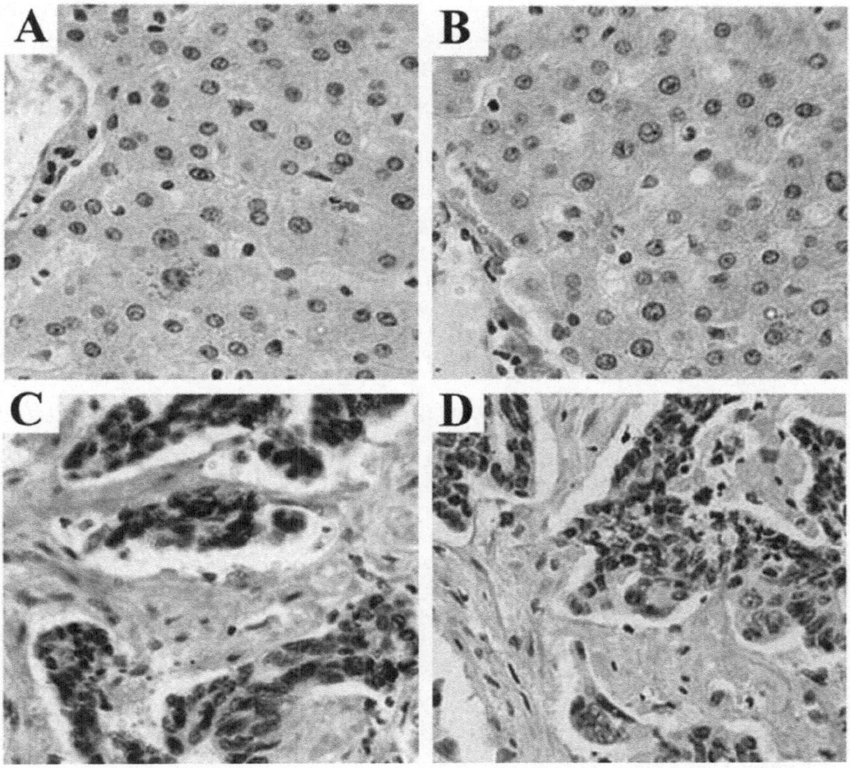

Figure 4.3. The absence of WIF-1 expression in adult human liver and its reactivation during cancer.

WIF-1 immunostaining in normal and diseased human adult liver. WIF-1 expression was not detected in normal adult liver parenchyma (A). In contrast WIF-1 expression was detected in adult liver with tumours containing SCM (C). Immunostaining controls stained negative, (B and D) thereby demonstrating specificity. Images were captured using a Leica FW4000 brightfield microscope at x20 magnification.

Color image of this figure appears in the color plate section at the end of the book.

controls (Fig. 4.3D). These results provide further evidence to suggest that the Wnt signalling equilibrium is vital in maintaining cellular stability *in vivo* and is regulated by stromal cell mesenchyme.

Conclusion

Although the use of stem cell based therapies is commonly discussed in regenerative medicine, another important application of this technology is in the generation of high fidelity human models. Human ESCs and iPSCs provide ideal models in which to study human biology and the influence of genotype at the molecular and cellular level. In order to gain a better understanding of Wnt signalling, during development and post natally, we employed primary human liver tissue and a stem cell derived tissue culture model. Both approaches demonstrate that the Wnt signalling equilibrium was important during human liver development and post-natal organ stability.

References

Arce L, NN Yokoyama and ML Waterman. 2006. Diversity of LEF/TCF action in development and disease. Oncogene. 25(57): 7492–7504.

Bossart P and KS Zaret. 1998. GATA transcription factors as potentiators of gut endoderm differentiation. Development. 125: 4909–4917.

Cirillo LA, FR Lin, I Cuestaand et al. 2002. Opening of compacted chromatin by early developmental transcription factors HNF3 (FoxA) and GATA-4. Mol Cell. 9(2): 279–289.

Clevers H. 2006. Wnt/beta-catenin signaling in development and disease. Cell. 127(3): 469–480.

Clotman F, P Jacquemin, N Plumb-Rudewiez et al. 2005. Control of liver cell fate decision by a gradient of TGF beta signaling modulated by Onecut transcription factors. Genes Dev. 19(16): 1849–1854.

Clotman F, VJ Lannoy, M Reber et al. 2002. The onecut transcription factor HNF6 is required for normal development of the biliary tract. Development. 129(8): 1819–1828.

Deng Y, B Yu, Q Chengand et al. 2010. Epigenetic silencing of WIF-1 in hepatocellular carcinomas. J Cancer Res Clin Oncol. 136(8): 1161–1167.

Ding Z, YB Qian, LX Zhu and QR Xiongand. 2009. Promoter methylation and mRNA expression of DKK-3 and WIF-1 in hepatocellular carcinoma. World J Gastroenterol. 15(21): 2595–2601.

Fletcher J, W Cui, K Samueland et al. 2008. The inhibitory role of stromal cell mesenchyme on human embryonic stem cell hepatocyte differentiation is overcome by Wnt3*a* treatment. Cloning Stem Cells. 10(3): 331–339.

Gadue P, TL Huber, PJ Paddison and GM Kellerand. 2006. Wnt and TGF-beta signaling are required for the induction of an *in vitro* model of primitive streak formation using embryonic stem cells. Proc Natl Acad Sci USA. 103(45): 16806–16811.

Haegel H, L Larue, M Ohsugiand et al. 1995. Lack of beta-catenin affects mouse development at gastrulation. Development. 121(11): 3529–3537.

Hay DC, J Fletcher, C Payne et al. 2008. Highly efficient differentiation of hESCs to functional hepatic endoderm requires Activin A and Wnt3a signaling. Proc Natl Acad Sci USA. 105(34): 12301–12306.

Hu J, A Dong, V Fernandez-Ruizand et al. 2009. Blockade of Wnt signaling inhibits angiogenesis and tumor growth in hepatocellular carcinoma. Cancer Res. 69(17): 6951–6959.

Jung J, M Zheng, M Goldfarb, and KS Zaretand. 1999. Initiation of mammalian liver development from endoderm by fibroblast growth factors. Science. 284(5422): 1998–2003.

Kawano Y and R Kypta. 2003. Secreted antagonists of the Wnt signalling pathway. J Cell Sci. 116(Pt 13): 2627–2634.

Kimelman D and W Xu. 2006. Beta-catenin destruction complex: insights and questions from a structural perspective. Oncogene. 25(57): 7482–7491.

Kinder SJ, TE Tsang, M Wakamiya et al. 2001. The organizer of the mouse gastrula is composed of a dynamic population of progenitor cells for the axial mesoderm. Development. 128(18): 3623–3634.

Koch A, A Waha, W Hartmannand et al. 2005. Elevated expression of Wnt antagonists is a common event in hepatoblastomas. Clin Cancer Res. 11(12): 4295–4304.

Kuhl M, LC Sheldahl, M Park et al. 2000. The Wnt/Ca2+ pathway: a new vertebrate Wnt signaling pathway takes shape. Trends Genet 16(7): 279–283.

Lawson KA, JJ Meneses and RA Pedersen. 1991. Clonal analysis of epiblast fate during germ layer formation in the mouse embryo. Development. 113(3): 891–911.

Lee AY, B He, L Youand et al. 2004. Expression of the secreted frizzled-related protein gene family is downregulated in human mesothelioma. Oncogene. 23(39): 6672–6676.

Liu P, M Wakamiya, MJ Sheaand et al. 1999. Requirement for Wnt3 in vertebrate axis formation. Nat Genet. 22(4): 361–365.

Logan CY and R Nusse. 2004. The Wnt signaling pathway in development and disease. Annu Rev Cell Dev Biol. 20: 781–810.

Lustig B, B Jerchow, M Sachsand et al. 2002. Negative feedback loop of Wnt signaling through upregulation of conductin/axin2 in colorectal and liver tumors. Mol Cell Biol. 22(4): 1184–1193.

Mazieres J, B He, L Youand et al. 2004. Wnt inhibitory factor-1 is silenced by promoter hypermethylation in human lung cancer. Cancer Res. 64(14): 4717–4720.

Micsenyi A, X Tan, T Sneddonand et al. 2004. Beta-catenin is temporally regulated during normal liver development. Gastroenterology. 126(4): 1134–1146.

Monga SP, HK Monga, X Tanand et al. 2003. Beta-catenin antisense studies in embryonic liver cultures: role in proliferation, apoptosis, and lineage specification. Gastroenterology. 124(1): 202–216.

Moon RT, JD Brown, JA Yang-Snyder et al. 1997. Structurally related receptors and antagonists compete for secreted Wnt ligands. Cell. 88(6): 725–728.

Nateri AS, B Spencer-Dene and A Behrens. 2005. Interaction of phosphorylated c-Jun with TCF4 regulates intestinal cancer development. Nature. 437(7056): 281–285.

Ober EA, H Verkade, HA Field and DY Stainierand . 2006. Mesodermal Wnt2b signalling positively regulates liver specification. Nature. 442(7103): 688–691.

Rhee CS, M Sen, D Lu et al. 2002. Wnt and frizzled receptors as potential targets for immunotherapy in head and neck squamous cell carcinomas. Oncogene. 21(43): 6598–6605.

Rossi JM, NR Dunn, BL Hogan and KS Zaretand. 2001. Distinct mesodermal signals, including BMPs from the septum transversum mesenchyme, are required in combination for hepatogenesis from the endoderm. Genes Dev. 15(15): 1998–2009.

Schier AF. 2003. Nodal signaling in vertebrate development. Annu Rev Cell Dev Biol. 19: 589–621.

Semenov MV, K Tamai BK Brott et al. 2001. Head inducer Dickkopf-1 is a ligand for Wnt coreceptor LRP6. Current Biology. 11: 951–961.

Shih YL, RY Shyu, CB Hsiehand et al. 2006. Promoter methylation of the secreted frizzled-related protein 1 gene SFRP1 is frequent in hepatocellular carcinoma. Cancer. 107(3): 579–590.

Shin D, CH Shin, J Tuckerand et al. 2007. Bmp and Fgf signaling are essential for liver specification in zebrafish. Development. 134(11): 2041–2050.

Shou J, F Ali-Osman, AS Multaniand et al. 2002. Human Dkk-1, a gene encoding a Wnt antagonist, responds to DNA damage and its overexpression sensitizes brain tumor cells to apoptosis following alkylation damage of DNA. Oncogene. 21(6): 878–889.

Sullivan GJ, DC Hay, IH Park et al. 2010. Generation of functional human hepatic endoderm from human induced pluripotent stem cells. Hepatology. 51(1): 329–335.

Suzuki H, DN Watkins, KW Jair et al. 2004. Epigenic inactivation of SFRP genes allows constitutive WNT signaling in colorectal cancer. Nature Genetics. 36: 417–422.

Tan X, J Behari, B Cieply et al. 2006. Conditional deletion of beta-catenin reveals its role in liver growth and regeneration. Gastroenterology. 131(5): 1561–1572.

Tang Y, AR Simoneau, WX Liaoand et al. 2009. WIF1, a Wnt pathway inhibitor, regulates SKP2 and c-myc expression leading to G1 arrest and growth inhibition of human invasive urinary bladder cancer cells. Mol Cancer Ther. 8(2): 458–468.

Thompson MD and SP Monga. 2007. WNT/beta-catenin signaling in liver health and disease. Hepatology. 45(5): 1298–1305.

Thomson JA, J Itskovitz-Eldor, SS Shapiro et al. 1998. Embryonic stem cell lines derived from human blastocysts. Science. 282(5391): 1145–1147.

Topol L, X Jiang, H Choiand et al. 2003. Wnt-5a inhibits the canonical Wnt pathway by promoting GSK-3-independent beta-catenin degradation. J Cell Biol. 162(5): 899–908.

Ueno K, M Hiura, Y Suehiro et al. 2008. Frizzled-7 as a potential therapeutic target in colorectal cancer. Neoplasia. 10(7): 697–705.

Urakami S, H Shiina, H Enokida et al. 2006a. Combination analysis of hypermethylated Wnt-antagonist family genes as a novel epigenetic biomarker panel for bladder cancer detection. Clin Cancer Res. 12(7 Pt 1): 2109–2116.

Urakami S, H Shiina, H Enokidaand et al. 2006b. Epigenetic inactivation of Wnt inhibitory factor-1 plays an important role in bladder cancer through aberrant canonical *Wnt/beta-*catenin signaling pathway. Clin Cancer Res. 12(2): 383–391.

Urakami S, H Shiina, H Enokidaand et al. 2006c. Wnt antagonist family genes as biomarkers for diagnosis, staging, and prognosis of renal cell carcinoma using tumor and serum DNA. Clin Cancer Res. 12(23): 6989–6997.

van Es JH, N Barker and H Clevers. 2003. You Wnt some, you lose some: oncogenes in the Wnt signaling pathway. Current Opinions in Genetic Development. 13: 28–33.

Weeraratna AT, Y Jiang, G Hostetter et al. 2002. Wnt5a signaling directly affects cell motility and invasion of metastatic melanoma. Cancer Cell. 1(3): 279–288.

Wei W, MS Chua, S Grepperand and SK So. 2009. Blockade of Wnt-1 signaling leads to anti-tumor effects in hepatocellular carcinoma cells. Mol Cancer. 8: 76.

Westfall TA, R Brimeyer, J Twedtand et al. 2003. Wnt-5/pipetail functions in vertebrate axis formation as a negative regulator of Wnt/beta-catenin activity. J Cell Biol. 162(5): 889–898.

Wirths O, A Waha, S Weggenand et al. 2003. Overexpression of human Dickkopf-1, an antagonist of wingless/WNT signaling, in human hepatoblastomas and Wilms' tumors. Lab Invest. 83(3): 429–434.

Wissmann C, PJ Wild, S Kaiserand et al. 2003. WIF1, a component of the Wnt pathway, is down-regulated in prostate, breast, lung, and bladder cancer. J Pathol. 201(2): 204–212.

Wu CH and R Nusse. 2002. Ligand receptor interactions in the Wnt signaling pathway in Drosophila. J Biol Chem. 277(44): 41762–41769.

Yamaguchi TP, S Takada, Y Yoshikawaand et al. 1999. T (Brachyury) is a direct target of Wnt3a during paraxial mesoderm specification. Genes Dev. 13(24): 3185–3190.

Zaret KS. 1999. Developmental competence of the gut endoderm: genetic potentiation by GATA and HNF3/fork head proteins. Developmental Biology. 209: 1–10.

Zhao R and SA Duncan. 2005. Embryonic development of the liver. Hepatology. 41(5): 956–967.

Zhao R, AJ Watt, J Liand et al. 2005. GATA6 is essential for embryonic development of the liver but dispensable for early heart formation. Mol Cell Biol. 25(7): 2622–2631.

Zhou Z, J Wang, X Han et al. 1998. Up-regulation of human secreted frizzled homolog in apoptosis and its down-regulation in breast tumors. International Journal of Cancer. 78: 95–99.

5

Applying Pluripotent Stem Cell Technology to Modelling Human Liver Disease

Nicholas R.F. Hannan,[1,a,*] *S. Tamir Rashid*[1,b] and *Ludovic Vallier*[1,c]

Introduction and interest in hIPSCs

The liver is the largest organ in the body and the principal site of a multitude of functions relating to the metabolism of carbohydrates, proteins, lipids, hormones and xenobiotic compounds, including prescription drugs. This broad spectrum of function exhibited by the liver is pivotal in supporting the functions of the other body organs and impairment in liver function often results in catastrophic effects for human health.

Recently the health burden resulting directly from liver related diseases such as viral hepatitis has dramatically increased (British Liver Trust web site http://www.britishlivertrust.org.uk). However the extent of this burden is significantly amplified when taking into consideration the indirect consequences generated by a wide spectrum of disease whose pathological processes are centred around the liver but whose ultimate pathological effects are seen within other organs, which occurs in the metabolic syndrome.

[1]Anne McLAren Laboratories for Regenerative Medicine, Department of Surgery, University of Cambridge, UK.
[a]Email: nrfh2@cam.ac.uk
[b]Email: str29@cam.ac.uk
[c]Email: lv225@cam.ac.uk
*Corresponding author

It is therefore imperative to address these growing challenges with relevant models of human disease to further our understanding of disease development and define novel therapeutic targets. Currently, the available models of liver disease may be broadly divided into *in vivo* (animal) or *in vitro* (cell based) models. Whilst some significant advances in the liver field have been made employing (largely rodent) animal models, it is clearly not possible to recapitulate the breadth or specificities of human physiology in such grossly different species. This probably explains to some extent the modest translational successes achieved to date employing such approaches. At the centre of this problem lies the inability to functionally replicate the unique repertoire of enzymes and proteins possessed by the human liver's principle cell type—the hepatocyte—a characteristic which is probably pivotal for delivering the human liver's broad ranging functional attributes. For example numerous attempts to study viral hepatitis biology and drug induced liver injury have been limited by the relatively poor correlation between human and animal hepatocyte functionality largely stemming from the inability to compensate for missing key receptors such as the Claudin receptor or missing key enzymes such as Cyp2C9.

The inability to replicate hepatocyte function has similarly impaired the utility of *in vitro* (human and non human) immortalized hepatocyte cell lines. As a result the only truly representative model shown to replicate human liver functionality *in vitro* is the primary human hepatocyte (Brandon et al. 2003) (Gomez-Lechon et al. 2003). Use of these cells however is not a viable long term solution for the field. Hepatocytes are usually taken from cadaveric organs making their availability and condition (viability and state of differentiation) highly variable. Availability of high-quality human hepatocytes is further hampered by the fact that they cannot be substantially expanded in tissue culture (Guillouzo and Guguen-Guillouzo 2008) and despite recent advances many hepatocyte specific disease processes still cannot be truly replicated using this model (Ploss et al. 2010).

The search for an alternative strategy by which to generate large quantities of high quality hepatocytes for disease modelling was given a major boost in 2007 when a landmark study from Shinya Yamanaka and colleagues opened up the possibility of deriving patient specific human induced plurpipotent stem cells (hIPSCs) by over expression of just a few transcription factors in somatic cells (Takahashi et al. 2007). Being pluripotent (and therefore embryonic stem cell like) meant these cells had the potential to be differentiated into any cell type of the human body in infinitely large quantities. In addition, being patient specific, allowed the possibility for the first time of modelling disease processes within a context of well characterized genetic and clinical profiles, therefore offering exciting new opportunities for cellular dissection of disease. Since 2007, some of the promises of hIPSCs have already been realized through successful *in vitro* modelling of various diseases (reviewed in (Rashid and Vallier 2010)) including inherited metabolic diseases of the liver (Rashid et al. 2010). Key to the ability of developing liver disease models with hIPSCs has been the translation of lessons derived from mammalian developmental biology to pluripotent cells in two dimensional culture as outlined below.

Developmental biology of the liver

The mammalian liver is derived from definitive endoderm and its embryonic origins, growth and maturation dynamics has been best described in the mouse. Following gastrulation, definitive endoderm migrates to form the primitive gut tube which consists of a single sheet of cells surrounded by mesoderm. Simultaneously, the primitive gut tube becomes regionalized along the dorsal-ventral and anterior posterior axes into ventral and dorsal foregut, midgut and hindgut domains. This patterning is induced by extracellular signals from the surrounding mesoderm including Wnt, FGF and BMP signal pathways. The effects of these signals can be observed at the molecular level with distinct expression patterns of genes such as SOX2 in the anterior foregut and CDX2 in the hindgut, while early PDX1 expression marks the midgut. It is from the foregut region however that the liver develops, along with thyroid, lung, the upper airways, billary system, stomach and ventral pancreas. The overlaying cardiac mesoderm (CM) and the septum transversum mesenchyme (STM) play an essential inductive role in liver development. They deliver essential BMP and FGF signalling gradients which in turn induce liver bud formation while patterning the remainder of the foregut in combination with other factors such as Wnts and Retinoic Acid (RA). Shortly after these inductive signals are given, the initially flat primitive gut closes to form the gut tube bringing together the cells committed to the hepatic lineage to form a liver bud.

Following BMP and FGF inductive signals, the liver bud itself becomes apparent as a proliferating outgrowth from the ventral foregut at approximately 8.5–9.0 DPC in the mouse, corresponding to approximately 3 wk in the human embryo. This process is marked by the expression of transcription factors such as HHEX, GATA4, GATA6, HNF6 and PROX1. At this stage, the newly committed hepatoblasts express AFP, TTR and ALB and also appear to be bi-potent by producing both hepatocytes and cholangiocytes. Interestingly, mice embryos which have both copies of these key regulatory genes "knocked out" display defects in liver organogenesis without losing liver bud formation. In embryos mutant for HEX for example, hepatoblasts fail to migrate into the septum transversum mesenchyme, whereas absence of HNF4a induces a wide range of liver defects including disregulation of genes controlling metabolism. Similarly, expression of

PROX1 is thought to be essential for hepatoblasts to delaminate and migrate into the STM.

Soon after the formation of the primary hepatic bud, the surrounding basement membrane is degraded and invaded by chords of hepatoblasts marked by GATA4 and HNF4a expression. Concurrently, angiogenesis and vasculogeniss proceed as the bud is permeated by angioblasts and endothelial cells. The bipotential hepatoblast population then differentiates into either hepatocytes or billiary epithelial cells mediated by the notch signal pathway. Finally, cells within the embryonic liver organise themselves into the complex hepatic architecture that is crucial for normal liver function.

How the myriad of different cells respond and differentiate in these final stages along with other questions pertaining to how the extracellular matrix is remodelled, how the hepatic vasculature and biliary systems become integrated and how intricate polarization of hepatocytes is established is only now beginning to be understood.

Protocols for hIPSC differentiation

A wide range of protocols have been published demonstrating the capacity of hESC's and hIPSC's to differentiate into hepatic-like cells *in vitro*. Most of these protocols focus on cell culture systems that mimic *in vitro* the main stages of embryonic development. These major stages include the pluripotent state, definitive endoderm (DE), foregut endoderm (FE), hepatic progenitors or hepatoblasts, and a matured hepatocyte.

All culture systems rely on high activin-A signalling to promote the initial differentiation step to DE. RPMI 1640 supplemented or not with B27 complement has been a common basal medium for the initial endoderm stage and past experience has shown that this basic media is as important as growth factors for the differentiation of hESCs. Using this approach, it is possible to obtain at least 80%–95% CXCR4 expressing DE cells (Teo et al. 2011). Genes commonly reported to be expressed following high activin treatment include, SOX17, Goosecoid, FOXA2, HHEX, and GATA4 demonstrating a high degree of similarity with DE and ADE *in vivo* (D'Amour et al. 2005), (Hay et al. 2008), (Rashid et al. 2010), (Zorn and Wells 2009).

Several different approaches have been reported to induce hepatic specification of the DE. Basal media include either RPMI1640 or KO-Serum Replacement containing medium and incorporate known inducers of the hepatic specification such as FGF and BMP growth factors. The most robust protocols result in 90% of the cells expressing AFP by the end of the induction period. In addition, the resulting cells also express the foetal hepatic markers AFP, Albumin, HNF4a, HNF6, A1AT, Cyp3A7, Cyp7A1, Cyp3A7, TAT, TTR while displaying functional activity characteristic of hepatocytes including glycogen storage, urea/ammonia cycle, cholesterol up take, detoxification associated with CYP expression and secretion of Alb and AAT. Nevertheless, these functions remain limited when compared to the activity of freshly isolated primary hepatocytes. Notably ALB, A1AT and Cyp3A4 expression are considerably lower when compared to adult expression. In effect, it is now commonly accepted that the hepatocytes generated from hPSCs maintain a foetal like identity. Accordingly, a broad number of groups have reported that whilst they are able to achieve quite efficient differentiation of embryonic stem cells into a hepatic-like cell type progress beyond foetal-functional characteristics has as yet to be achieved (Touboul et al. 2009, Hay et al. 2008, Rashid et al. 2010, Duan et al. 2010).

OSM and HGF are systematically employed to induce maturation and to maintain proliferation of hepatocyte-like cells generated from hPSCs. While cells do initially proliferate and begin to mature after 10–15 d in most of the culture systems, they also begin to show signs of senescence and deregulation of cellular metabolism as large lipid vacuoles accumulate in the cytoplasm. The current limitations of the primary hepatocyte culture system of dedifferentiation, growth inhibition and functional decline have also been observed with hPSCs derived hepatocytes and is likely to result from suboptimal culture conditions that do not mimic the physiological condition of the liver. Synthetic growth surfaces have been used to optimize and improve hepatocyte growth and functionality, however without appropriate and optimal metabolic and growth requirements cells inevitably senesce and lose functionality over time (Hay et al. 2011). Thus, the capacity of these cells to be used for therapeutic applications will need to be carefully assessed. However it has already been demonstrated that once these immature or foetal hepatic progenitors are injected into the liver, they do have the capacity to integrate, survive, and function in a manner analogous to adult hepatocytes. Furthermore these cells still provide a valuable tool for modelling human metabolic liver disease.

Examples of liver diseases modelled with hISPCs

The current potential of hIPSC based disease modelling systems is best illustrated by consideration of the most commonly occurring inherited metabolic liver disease, α_1-antitrypsin deficiency. Previous studies demonstrated that a homozygously inherited single base pair mutation in exon V of the SerpinA1 gene results in protein misfolding and entrapment of mutant (Z) α_1-antitrypsin polymers within the endoplasmic reticulum (ER) of patient hepatocytes which in turn results in downstream liver injury and hepatic disease (Gooptu and Lomas 2009). Several key questions remain unanswered in this field, in particular, how does accumulation of Z $\alpha_1$antitrypsin polymers ultimately cause hepatocyte cell death? And why is there a variation in clinical phenotype observation between individuals of the same genetic background?

Underpinning these questions, also relevant to a host of other protein misfolding disorders, is the role that protein degradation pathways play in human hepatocyte biology. Therefore use of previously employed *in vitro* models such as hepatic (Teckman et al. 2001) and extrahepatic mammalian cell lines (Qu et al. 1996) which rely on artificially high levels of induced mutant protein expression, is unlikely to bring further clarity to the subtle processes underpinning hepatocyte protein handling *in vivo*. Instead a new approach, more closely replicating human physiology but importantly, also within the context of a precisely defined genetic background has long been required.

By generating patient specific α_1-antitrypsin hIPSC derived hepatocytes capable of conserving disease specific protein polymerisation and ER entrapment, we showed that such an *in vitro* cellular platform is now available (Rashid et al. 2010). Further studies must now be performed to see whether indeed the new hIPSC technology can help to unravel the outstanding issues alluded to previously.

Two other liver diseases have also been modelled with hIPSC technology and reinforce the demonstration that downstream effects of protein dysfunction in hepatocytes may be conserved within a patient specific manner. Firstly by replicating the lack of LDL receptor mediated uptake of LDL lipid into hepatocytes as occurs in Familial Hypercholesterolemia, we successfully showed that this technology is equipped to provide a holistic impression of the hepatocyte specific native processes governing receptor dysfunction, from nuclear synthesis to trafficking through the ER and ultimately recycling at the cell membrane. This suggests that employing a hIPSC platform could be used for follow up studies into a broad range of liver related receptoropathies.

Finally by demonstrating the conservation of several key facets of the Glycogen storage disease type 1a (GSD-1a) cellular phenotype and the hepatocyte specific glucagon induced upregulation of key downstream gluconeogenic enzymes and a secreted IGF1 binding protein, we provided a "proof of concept" that similar studies to investigate other more common metabolic disorders could be feasible using an iPS platform. Of note with this last example, however, was the observation that insulin-induced enzyme responses were not adequately demonstrated. This highlights one of the unresolved challenges facing hIPSC derived cell models for modelling liver diseases, the production of fully mature cells.

Future challenges

The possibility of producing large quantities of high fidelity hepatocytes from hPSCs could represent a major development not only for regenerative medicine but also for drug development. Nevertheless, major challenges need to be resolved before envisaging the use of such hPSCs derived hepatocytes *in vitro* or *in vivo*.

The first challenge is the difficulty posed by the variable capacity of hPSCs to differentiate into specific lineages. Indeed, several reports have shown that hESCs or hIPSCs can display some preferences when it comes to differentiating into the primary germ layers endoderm, mesoderm end neuroectoderm. As a consequence, specific hPSCs lines can not differentiate efficiently into DE and thus into hepatocytes. This variability between lines represents a major limitation to production of hepatocytes from a broad number of individuals of diverse genetic backgrounds which is a prerequisite for the development of tailored drug/toxicology screens. In addition, variability between hIPSCs could represent a major challenge for their use in the context of personal cell based therapy.

The cause of this variability between hPSCs remains unclear, though epigenetic anomalies either induced by *in vitro* cell culture or by reprogramming remain the usual suspects. However, studies in the mouse have clearly showed that genetic background can have a major influence in the capacity of mESCs to differentiate *in vitro*. Therefore, it is likely that a combination of epigenetic and genetic factors could define the capacity of differentiation of hPSCs. Recent reports have shown that variability could be solved by adapting culture conditions for each line employed (Boulting et al. 2011). In addition, gene expression profiling can now be used to determine the differentiation capacity of each line facilitating the identification of optimal culture conditions for specific lines. However, such approaches may prove difficult to apply if variability between lines also occurs at each differentiation steps. Therefore, alternative approaches are urgently needed and an advantageous solution could be to genetically and or epigenetically engineer hPSCs to make them more prone to differentiate into a specific germ layer. This could be achieved by overexpression or knock down of specific transcription factors responsible for controlling cell fate decisions (Zhang et al. 2010) in combination with the use of novel and defined cell substrates.

The difficulty of generating fully mature, functional cell types from hPSCs also represents a key challenge to delivering their highly anticipated promise. Importantly, this issue is not restricted to hepatocytes. Haematopoeitic Stem Cells (HSCs) generated from hPSCs resemble stem cells of the yolk sac and differentiate into erythrocytes expressing foetal globin. Cardiomyocytes display contractile capacity more akin to immature cells whilst neurons generate axon potentials some way removed from those observed in primary adult cells. This limitation could prove problematic for *in vitro* analyses requiring fully mature cell types such as drug screening. Similarly, the use of fully differentiated cells may be preferable in the context of cell therapy since achieving a therapeutic effect should prove more efficient but also more safe with adult like cells than if progenitors were used which could over proliferate and/or dedifferentiate more easily.

Thus development of novel approaches to improve maturity of cells generated from hPSCs is a major objective in the stem cell field. The complexity and the time frame of tissue maturation occurring during normal human development represent the major hurdles. Indeed, the information concerning the mechanisms controlling maturation of foetal hepatocytes remain limited by ethical and technical difficulties associated with the use of human foetal tissues. Furthermore, hepatocytes seemingly become fully functional only at or after birth implying that hPSC derived hepatocytes might require several months of differentiation *in vitro* to become fully functional. For all these reasons, alternative approaches to direct differentiation are currently being explored including the possibility of transdifferentiating somatic cells (Huang et al. 2011, Sekiya and Suzuki 2011). Despite promising initial results, the real interest of such approaches for generating functional cells fully compatible with clinical applications on a large scale remains to be careful examined.

Despites all these difficulties, hESCs have recently been given the go ahead to be examined within the context of clinical trials for treatment of spinal cord injury and blindness (Alper 2009). Additional clinical trials are now in preparation and will focus on cell types such as cardiomyocytes and keratinocytes suggesting that clinical applications of hPSCs will not be limited to one particular organ or disease. Hepatcoyte based therapy could be the next application for hESCs. Importantly, we can now produce near homogenous populations of foetal hepatocytes in culture conditions compatible for GMP production and thus the practical challenges usually related to this kind of therapy could easily be overcome. Nevertheless, two major aspects remain to be addressed before any clinical trial in man.

Safety of hepatocytes generated from hPSCs needs to be demonstrated *in vivo* not only for the tumorigenic risk associated with pluripotent stem cells but also for the unknown dangers presented by the transplantation of heterogeneous populations of cells into the liver. Indeed, generation of pure populations of hepatocytes from hPSCs remains impossible to date. Both these safety issues could be avoided by purifying hepatocytes just before use. However, this option might be technically challenging since cell sorting methods like FACS are not compatible with the large scale production necessary for cell therapy and could potentially diminish the functionality of the purified cells. Thus, new technical development would be required to ensure the safety of hepatocytes generated from hPSCs. In addition, even if cell sorting were possible it would not avoid the need for extensive testing in animal models to demonstrate absence of tumour formation or overgrowth of progenitors.

Extended proof of principle is also needed to justify the use of hPSCs derived hepatocytes in man. For that, comprehensive studies are urgently needed to show that hepatocytes generated from hESCs or hIPSCs can colonize the liver of animal models of liver failure while maintaining functionality. UPA-SCID and FRG mice have been extensively used with primary hepatocytes and thus represent the best models available so far to test the clinical potential of hPSCs derived hepatocytes. However, these mice are more relevant for modelling inherited metabolic disorders but may not be appropriate models of liver failure induced by other aetiologies. In addition, the mouse system has a limited relevance in terms of size, metabolism, physiology, life duration and immunology. Non human primates or pigs could represent interesting alternatives but new disease models will need to be developed in these species and this could take a number of years and considerable expense. In the interim, the possibility of generating humanized mice could help address a number of these questions, especially the possible immunogeneicity of hepatocytes derived from hPSCs. Therefore, the combination of extensive testing in existent and new animal models would be a necessary for demonstrating the possible application of hPSC derived hepatocytes for cell therapy of liver disease.

To conclude, the incredible speed of progress achieved to date in generation of hepatocytes from human pluripotent stem cells makes the prospect of their *in vivo* applications a fast approaching reality. It is now time for basic scientists to work diligently alongside clinicians in a final push to realize the last steps of optimization that will lead to delivery of the exciting full spectrum of clinical benefit for patients offered by hPSCs.

References

Alper J. 2009. Geron gets green light for human trial of ES cell-derived product. Nat Biotech. 27(3): 213–214.

Boulting GL, E Kiskinis, GF Croft et al. 2011. A functionally characterized test set of human induced pluripotent stem cells. Nat Biotechnol. 29(3): 279–286.

Brandon EF, CD Raap, I Meijerman et al. 2003. An update on *in vitro* test methods in human hepatic drug biotransformation research: pros and cons. Toxicol Appl Pharmacol. 189: 233–246.

D'Amour KA, AD Agulnick, S Eliazer et al. 2005. Efficient differentiation of human embryonic stem cells to definitive endoderm. Nat Biotech. 23(12): 1534–1541.

Duan Y, X Ma, C Wang et al. 2010. Differentiation and Characterization of Metabolically Functioning Hepatocytes from Human Embryonic Stem Cells. Stem Cells. 28(4): 674–686.

Gomez-Lechon MJ, MT Donato, JV Castell et al. 2003. Human hepatocytes as a tool for studying toxicity and drug metabolism. Curr Drug Metab. 4: 292–312.

Gooptu B and DA Lomas. 2009. Conformational Pathology of the Serpins: Themes, Variations, and Therapeutic Strategies. Annual Review of Biochemistry. 78(1): 147–176.

Guillouzo A and C Guguen-Guillouzo. 2008. Evolving concepts in liver tissue modeling and implications for *in vitro* toxicology. Expert Opinion on Drug Metabolism & Toxicology. 4(10): 1279–1294.

Hay DC, D Zhao, J Fletcher et al. 2008. Efficient differentiation of hepatocytes from human embryonic stem cells exhibiting markers recapitulating liver development *in vivo*. Stem Cells. 26(4): 894–902.

Hay DC, S Pernagallo, JJ Diaz-Mochon et al. 2011. Unbiased screening of polymer libraries to define novel substrates for functional hepatocytes with inducible drug metabolism. Stem Cell Research. 6(2): 92–102.

Huang P, Z He, S Ji et al. 2011. Induction of functional hepatocyte-like cells from mouse fibroblasts by defined factors. Nature. Advance online publication.

Ploss A, SR Khetani, CT Jones et al. 2010. Persistent hepatitis C virus infection in microscale primary human hepatocyte cultures. Proceedings of the National Academy of Sciences. 107(7): 3141–3145.

Qu D, JH Teckman, S Omura et al. 1996. Degradation of a Mutant Secretory Protein, alpha 1-Antitrypsin Z, in the Endoplasmic Reticulum Requires Proteasome Activity. J Biol Chem. 271(37): 22791–22795.

Rashid ST and L Vallier. 2010. Induced pluripotent stem cells? alchemist's tale or clinical reality? Expert Reviews in Molecular Medicine. 12(-1): p. NA.

Rashid ST, S Corbineau, NRF Hannan et al. 2010. Modeling inherited metabolic disorders of the liver using human induced pluripotent stem cells. The Journal of Clinical Investigation. 120(9): 3127–3136.

Sekiya S and A Suzuki. 2011. Direct conversion of mouse fibroblasts to hepatocyte-like cells by defined factors. Nature. Advance online publication.

Takahashi K, K Tanabe, M Ohnuki et al. 2007. Induction of Pluripotent Stem Cells from Adult Human Fibroblasts by Defined Factors. Cell. 131(5): 861–872.

Teckman JH, J Burrows, T Hidvegi et al. 2001. The Proteasome Participates in Degradation of Mutant Î±1-Antitrypsin Z in the Endoplasmic Reticulum of Hepatoma-derived Hepatocytes. Journal of Biological Chemistry. 276(48): 44865–44872.

Teo AKK, SJ Arnold, MWB ans Trotter et al. 2011. Pluripotency factors regulate definitive endoderm specification through eomesodermin. Genes & Development. 25(3): 238–250.

Touboul T, NRF Hannan, S Corbineau et al. 2009. Generation of functional hepatocytes from human embryonic stem cells under chemically defined conditions which recapitulate liver development. Hepatology. 9999(999A): p. NA.

Zhang X, CT Huang, J Chen et al. 2010. Pax6 Is a Human Neuroectoderm Cell Fate Determinant. Cell stem cell. 7(1): 90–100.

Zorn AM and JM Wells. 2009. Vertebrate Endoderm Development and Organ Formation. Annual Review of Cell and Developmental Biology. 25(1): 221–251.

6

Applying Stem Cell Technology to High Through-put Drug Screening

Mia Emgård[1, 2,*] and *Petter Björquist*[1]

Introduction

The pharmaceutical industry is experiencing increasing costs for development of new drugs, partly due to the comprehensive safety assessment process and high drug attrition rates. Toxicity testing is therefore an essential part of the evaluation of potential new drugs. The rate of success has been estimated to be as low as around 11% (Kola 2008), with the majority of failures occurring in late Phase III of clinical trials, at the time of registration or even after market entrance. This has led to discussions and reports from both the U.S. Food and Drug Administration (FDA; http://www.fda.gov) and the European Medicines Agency (EMA; http://www.ema.europa.eu). Additionally, there is a growing need for alternative safety assessment assays with the current pressure to reduce animal testing of new cosmetics- and chemical products. In fact, over the next decade, the European Union joint program regulation on Registration, Evaluation, Authorisation and restriction of CHemicals (REACH) aims to

[1] Cellectis Stem Cells Cellartis AB, Arvid Wallgrens Backe 20, 41346 Göteborg, Sweden; Email: mia.emgard@cellectis.com, petter.bjorquist@cellectis.com

[2]Dept of Neurobiology, Care Sciences and Society, Karolinska Institutet, Stockholm, Sweden.

**Correspondance address*: Mia Emgård, Cellartis AB, Arvid Wallgrens Backe 20, 41346 Göteborg, Sweden. Email: mia.emgard@cellectis.com

assess the potential toxic effect of all chemicals sold in Europe. This is the largest investment ever of consumer safety in Europe, with a focus on the development of high through-put alternatives (Hartung and Rovida 2009, Rovida and Hartung 2009). *In vitro* safety assessments available today in drug discovery and evaluation of environmental stress on human health are either based on non-human material or limited sources of human tissue samples. Unfortunately, big inter-experimental variations particularly when using human material, negatively impacts the process, and thus, there is a great need to evaluate alternative options, which employ quality assured hepatocytes from a stable genotype. Pluripotent stem cells offer unique possibilities and have therefore been discussed as an alternative for toxicity studies (Améen et al. 2008, Wobus and Löser 2011). Qualities such as an unlimited supply of cells with natural high proliferation rate, possibilities to differentiate into all cell types in the body and being cell lines derived from different polymorphic variants and from specific patients are some of many advantages of using pluripotent stem cells for toxicity testing (Ebert and Svendsen 2010, Laustriat et al. 2010, Sartipy and Björquist 2011). Although there are several novel and interesting technologies for toxicity testing of various stem cells and their derivates, this chapter will focus on human pluripotent stem cell (hPSC) derived hepatocytes and their possible role in high through-put *in vitro* toxicity assays.

Drug induced liver injury

Drug induced liver injury (DILI) is the single most frequent reason for late attrition and removal of approved drugs from the market, and has been estimated to contribute to more than 50% of the failed drugs (Abboud and Kaplowitz 2007, Suzuki et al. 2010). DILI can be induced either by direct effect of a certain drug or chemical, or it can be due to metabolites generated by biotransformation pathways. The toxic compounds can have an effect on cell viability or at the genetic level, resulting in aberrant gene expression of for example drug metabolizing enzymes. In this scenario altered drug metabolism could result in accumulation of toxic metabolites and alterations of biochemical processes (Gómez-Lechón et al. 2010).

Human liver drug metabolism is generally divided into three phases: Phase I, II and III. Phase I metabolism typically involve oxidative reactions with cytochrome P450 monooxygenases (CYPs), including CYP1A2, CYP2A6, CYP2B6, CYP2Cs, CYP2D6, CYP2E1, CYP3A4, and CYP3A5 (Phillips and Crook 2010). Phase II reactions are usually detoxicative in nature, and involve conjugation of Phase I metabolites, with, e.g., carboxyl-, hydroxyl-, amino-, and sulfhydryl groups. The resulting conjugates are usually inactive, unlike the commonly active and often harmful metabolites generated during Phase I reactions. In Phase III, the metabolites and conjugates are transported out of the cells, and can affect other organs. Thus, an ideal hepatocyte for *in vitro* toxicity testing, should exhibit typical hepatic functions, such as stable expression of enzymes involved in Phase I and II of drug metabolism and with functional Phase III transporters. Such cells should optimally be sourced from a defined genetic background available over time, and be able to be mass-produced at high homogeneity, with small batch-to-batch variations.

Current *in vitro* hepatotoxicity models

In preclinical safety assessment studies, animal *in vivo* experiments are still commonly utilized. Toxicity studies using cell cultures not only decrease animal suffering and the high costs associated with animal experiments, but could also be scaled up and thereby increase the throughput in an industrial setting. Such frontloading of testing would help early identification of problematic compounds. In addition, data generated from toxicity tests using efficient human cell models would, logically, be much more predictive of human hepatotoxicity.

Cultured human primary hepatocytes are today the golden standard of functional liver cells, as they exhibit a full functional and inducible hepatic drug-metabolizing enzyme arsenal (Gómez-Lechón et al. 2010). Unfortunately, these cells have limitations since they lose their tissue specific characteristics such as enzyme activity and transporter functions within hours, are not available from the same genetic background (individual) over time, and are hampered by large variations between batches (Sartipy et al. 2007, Bickle 2010). The primary human hepatocytes are most often derived from low quality fragments of human liver tissue (Wobus and Löser 2011). Alternatively, hepatocyte-like cells from the human immortalized hepatoma cell lines HepG2 and HepaRG, can be used for toxicity testing (Guguen-Guillouzo and Guillouzo 2010). However, these cells exhibit low expression of crucial hepatocyte-specific genes, have only low enzyme activities of the CYP450 family members critical for Phase I metabolism, require long term culture before peak activity, are inherently limited in human genetic diversity, and are karyotypically abnormal (Greenhough et al. 2010, Wobus and Löser 2011).

Pluripotent stem cells in drug discovery

During the last decade, hPSC, being human embryonic stem cells (hESC) and human induced pluripotent stem cells (hiPSC), have received increasing appreciation for their use in a wide range of experimental settings and clinical applications (Améen et al. 2008, Liras 2010). Initially, there was a great focus on clinical cell replacement therapies, but with innovative research the scientific community has gained knowledge and experiences that have broaden our view of possible applications for hPSC (Wobus and Löser 2011). Particularly, the use of hPSC will most probably be of increasing importance for the specificity and efficiency of *in vitro* drug screening and toxicity testing in the future (Cai et al. 2007, Duan et al. 2007, Agarwal et al. 2008, Hay et al. 2008a, Hay et al. 2008b, Basma et al. 2009, Touboul et al. 2010, Hay et al. 2011).

For differentiation of hESC, attempts are made to mimic the *in vivo* conditions of hepatocyte differentiation. It is previously known that hepatocytes are highly dependent on cell-to-cell/matrix interactions, both in the developing and adult liver (Si-Tayeb et al. 2010a). Indeed, it is commonly known that as soon as primary hepatocytes are removed from the *in vivo* three dimensional structures and cultured *in vitro*, the cells will lose most of their tissue specific characteristics (Hewitt et al. 2007). To develop a fully functional hepatocyte from undifferentiated stem cells, it appears that the cells need to differentiate *in vitro* as they do *in vivo*, i.e., through steps of definitive endoderm, hepatoblasts, and finally to mature adult hepatocyte (Hay et al. 2008b, Brolén et al. 2010). Genes characteristic for each stage of the developing liver can be used as markers to evaluate the *in vitro* differentiation process and its quality control. Several research groups have reported on the induction of hepatocyte specific genes, and albumin positive cells with clear hepatocyte morphology (Hay et al. 2007, Basma et al. 2009, Miki et al. 2011). However, so far, only a few studies have been published on functional capacity of hESC derived hepatocyte-like cells, similar or close to that of freshly isolated human primary hepatocytes (Duan et al. 2010, Hay et al. 2011) (Fig. 6.1). In a further attempt to resemble the *in vivo* situation, co-cultures with liver epithelial and/or vascular endothelial cells have been discussed (Guguen-Guillouzo and Guillouzo 2010). The liver has precise tissue architecture *in vivo* and consists of several different cell types, which all contribute to the metabolism or resolution of a particular chemical exposure (Trosko and Chang 2010a). Hence, if it would be feasible to culture hepatocytes together with their natural cell neighbours, the predictive power of the hepatotoxicity assay would most probably be improved. For drug development or safety assessment analysis, long term cultures with sustained phenotypes would be desirable, as detection of drug effects might require the application over an extended period of time or reiterated exposure. However, till date, very few studies have reported on extended culture times for more than 20–30 d *in vitro* (Shiraki et al. 2008). A recent study by Hay and co-workers reported on a dramatic effect on cell viability, tissue specific gene expression, and drug-inducible metabolism of hESC derived hepatic endoderm cultured on polyurethane matrix (Hay et al. 2011). In addition, it was reported that the long-term

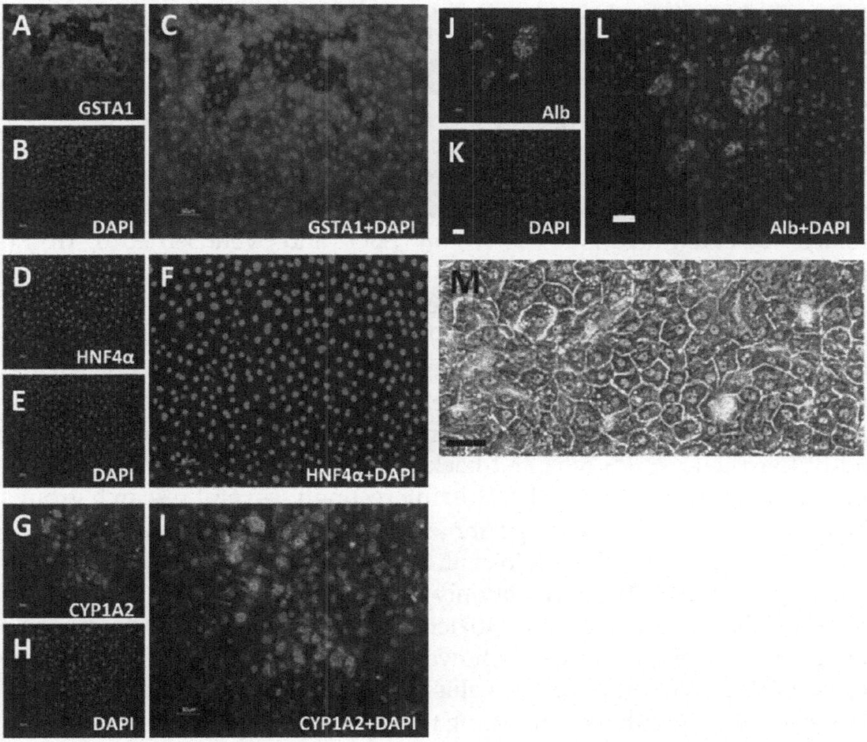

Figure 6.1. Microphotographs of functional human embryonic stem cells derived hepatocyte-like cells, cultured in a feeder-free system. A: Hepatocyte-like cells positive for Glutathione S-transferase 1 (GSTA1), a Phase II enzyme crucial for drug metabolism. B: Nuclei stained with DAPI. C: Merged photograph of A and B. D: HNF4α positive cells. The transcription factor HNF4α is important during differentiation to hepatocytes, and is used as a marker for differentiated hepatocytes. E: Nuclei stained with DAPI. F: Merged photograph of D and E. G: Cells positive for CYP1A2, a Phase I drug metabolizing enzyme and a marker of functional hepatocytes. H: Nuclei stained with DAPI. F: Merged photograph of G and H. J: Albumin staining, showing production of albumin in hepatocyte-like cells. K: Nuclei stained with DAPI. L: Merged photograph of J and K. M: Bright-field microscopic image displaying typical hepatocyte morphology of polygonal cells with bright nucleus and dark cytoplasm, large cytoplasm to nucleus ratio, nuclei with one to two prominent nucleoli, and numerous bi- or poly-nucleated cells. Scale bar= 50μM. (Courtesy of Cellartis AB).

Color image of this figure appears in the color plate section at the end of the book.

hepatocyte function and growth was supported by this clinically approved bioartificial liver matrix.

A proposal of standardized methods to assess stem/progenitor cells or hepatic lineage-committed progeny has been made by a group consisting of a basic biologist and clinical hepatologists (Sancho-Bru et al. 2009). Although the purpose of these standardized criteria aimed for clinical cell therapy, they could possibly also be applicable for characterization of hepatocytes for *in vitro* toxicity applications.

Ever since Takahashi and Yamanaka published the first report in 2006 on successful reprogramming of adult somatic cells to cells with embryonic stem cell like characteristics, there has been an ebullient research activity on possible applications of these so called induced pluripotent stem cells (Takahashi and Yamanaka 2006). The safety assessment and toxicity research field is no exception (Ebert and Svendsen 2010, Trosko and Chang 2010a). Hepatocyte-like cells generated from hiPSC has been shown by several groups (Ghodsizadeh et al. 2010, Si-Tayeb et al. 2010b, Sullivan et al. 2010, Rashid et al. 2010, reviewed in Asgari et al. 2010), and although the origin of the donor cells and the reprogramming factors were different, the hepatocyte-like cells expressed genes in a liver-specific manner and exhibited certain hepatocyte functions. In fact, a full loop, from adult liver cells to iPS cells and back to mature hepatocytes has recently been accomplished (Liu et al. 2010). Interestingly, several research groups have reported that the reprogrammed cells seem to have an "epigenetic memory" (Kim et al. 2010, Polo et al. 2010). Consequently, the somatic cell "memory" of the iPSC may not only affect the differentiation preference, but may also have an impact on toxicity studies (Hu et al. 2010, Wobus and Löser 2011). These findings are however debated and more studies have to be performed before any conclusions can be drawn (Meissner 2010). For example, it would be interesting to compare iPSC derived hepatocytes originating from liver or fibroblasts from the same individual. Nevertheless, recent studies that have compared hESC and hiPSC have come to the conclusion that the difference between hESC and hiPSC is less than what has been observed between different hESC-lines (Cellartis-unpublished observation). Individual differences between cell lines can however be a great advantage, as there is a big variation between human individuals in the level of drug metabolism (Rashid et al. 2010). Thus, polymorphic variants of hiPSC lines would allow the investigation of individual drug metabolism and related toxic effects, studies not accessible with other techniques (Greenhough et al. 2010). Furthermore, the prospect of using hiPSC specific for disease or groups of drug-responders, e.g., DILI, would imply tremendous possibilities for toxicity studies (Hochedlinger and Plath 2009, Liu et al 2010, Kang 2010).

Cell-based assays for high through-put drug discovery and predictive toxicity

Although we can gain a great deal of information from animal toxicity studies, the results cannot always be extrapolated to predict human drug toxicity (Olson et al. 1998, 2000). Screening of new chemical entities early in drug discovery is dependent on predictive safety biomarkers that are detectable and quantifiable prior to the onset of irreversible tissue damage. There are several different methods available to evaluate alterations in gene expression, post-translational modifications and cell damage, e.g., high-content imaging, genetic profiling with RTqPCR or microarrays, proteomics, metabonomics or epigenetics (Gómez-Lechón et al. 2010). However, not all of the mentioned techniques are suitable for evaluation at a high through-put industrial scale.

Approaches to measure toxicity endpoints include cell membrane integrity, mitochondrial and lysosomal function, as well as the content of cellular metabolites (O'Brien et al. 2006, Gómez-Lechón et al. 2010) (Fig. 6.2). There are numerous different methods to evaluate cytotoxicity in high through-put assays, including signs of protease activities, release of proteins from disrupted cellular membranes, chromatin condensation or total

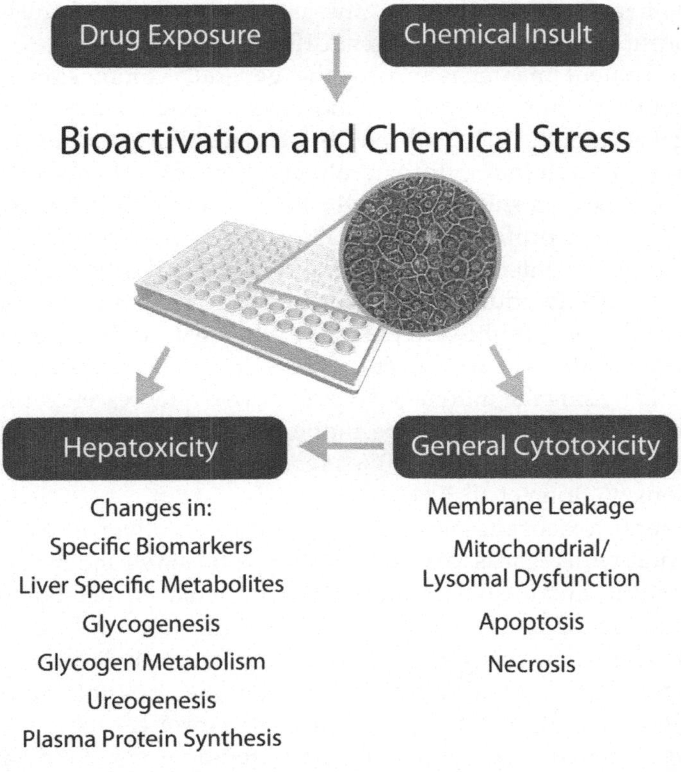

Figure 6.2. Toxicity evaluations of drugs or chemicals *in vitro* can be initiated by assessing classical cytotoxic endpoints, such as membrane leakage, mitochondrial or lysosomal dysfunction, apoptosis and necrosis. However, for more hepatotoxic specific evaluations, alterations in other parameters can be studied, such as, e.g., specific biomarkers, liver specific metabolites, glycogenesis, glycogen metabolism, ureogenesis and plasma protein synthesis.

cellular disintegration (Gómez-Lechón et al. 2010). However, cytotoxicity measures lethal events often at a late stage of toxicity, and is therefore considered as a blunt instrument to evaluate toxic alternations at sublethal toxicity doses (O'Brien et al. 2006). Thus, to correctly predict liver specific toxins, tissue specific parameters should be assessed (Gómez-Lechón et al. 2010). To design a useful predictive hepatotoxicity model, it is important to evaluate changes in biomarkers and liver-specific metabolites that occur prior to the initiation of the cell death machinery. Impaired liver function, without cellular degeneration, can be of significance not only for the organ, but in most cases also for the whole body. For liver toxicity specific read outs, hepatic function can for example be assessed by evaluating metabolic parameters characteristic for hepatocytes, such as glucogenesis, glycogen metabolism, ureagenesis, and plasma protein synthesis (Gómez-Lechón et al. 2010) (Fig. 6.2). Moreover, rapid release of liver specific protein biomarkers such as alanine- and aspartyl aminotransferase and bilirubin, would enhance the sensitivity of the test, ideally for an industrial high through-put safety study (Amacher 2010, Shi et al. 2010).

High-content imaging is an assay that generally combines engineering, biological endpoints, informatics and data mining. Using high-content imaging it is possible , in high through-put, to monitor multiple parameters indicative of sublethal cytotoxic effects, both by morphological and biochemical mechanisms (Barbaric et al. 2010, Bickle 2010). For example, it is possible to get a profile of compound-specific "stress" responses, such as elevated tubulin content, mitochondrial swelling and actin reorganization (Amacher 2010). In addition, high-content imaging can be performed on a single-cell level in multi-well plate formats, ideal for tissue specific high through-put analysis (Abraham et al. 2008, Lou and Liang 2011). It has been reported that, using this automated quantitative epifluorescence microscopy technique with analysis of liver specific biomarkers, the specific hepatoxicity read out can be increased by 80–90%, as compared to traditional cytotoxicity evaluations (O'Brien et al. 2006).

The term toxicogenomics refers to studies on adverse effects of test compounds, with regards to the relationship between structure and activity of the genome, and includes studies using proteomics, metabolomics and transcriptomics (Gómez-Lechón et al. 2010). Proteomics and metabolomics are techniques where changes in specific protein or metabolite configurations and contents are studied. Epigenetics is another possible approach to study alterations due to toxic exposure. The term can be described as stable, yet reversible, heritable molecular alterations generating a specific phenotype, without genotype changes (Meissner 2010, Trosko and Chang 2010a). Exposure of chemicals or drugs may cause DNA methylation or demethylation, as well as histone modifications, leading to loss or gain of function of certain proteins (Meissner 2010, Trosko and Chang 2010a).

These modifications can be monitored, classified into level of toxic damage and used in toxicity prediction models. The above mentioned techniques are based on mass spectrometry, which can give specific information on pathological processes or responses to toxic exposure (Amacher 2010). However, even though proteomics, metabolomics and epigenetics are ideal for biomarker identification, the techniques are today not suitable for high through-put toxicity analyses (Gómez-Lechón et al. 2010). On the contrary, transcriptomic evaluations can be performed in high throughput. Studies on alternations in gene expression using microarray analyzing techniques is currently considered both relevant and reproducible, and has become the most frequently used approaches in toxicogenomics (Guguen-Guillouzo and Guillouzo 2010). With DNA or RNA microarrays, it is possible to simultaneously monitor modifications of thousands of genes, with the prospect to distinguish between different subtypes of hepatoxicity at gene expression level. Finally, biostatistical analysis of data, generated after exposure of a set of compounds with known toxic mechanisms, can uncover unique gene expression profiles (fingerprints or signatures) and can be possibly, used as a database, to identify the potential toxic effect of new chemicals or drugs (Waters and Fostel 2004, Torsko and Upham 2010b, Kang and Trosko 2011, Lasher et al. 2011, Yildirimman et al. 2011).

Essential steps in a high through-put hepatoxicity assessment

One of the great advantages with stem cells is their capacity to proliferate with sustained genetic profile and differentiation capacity, without cancer cell heritage or introduction of immortalizing genes (Lou and Liang 2011). To optimize scale up protocols for high through-put cell culture assays in a well-regulated environment, requires cutting edge biotechnology (Upadhaya and Selvaganapathy 2009, Serra et al. 2010, Wu et al. 2010). The propagation of undifferentiated stem cells to large numbers in a cost effective manner will require novel bioreactor technology, with the aim to allow cell proliferation without compromising viability or pluripotency capacity. As examples, stirred tanks with or without microbeads, multilayer flasks, disposable cell bags, and roller flasks have been explored (King and Miller 2007, Kehoe et al. 2010, Serra et al. 2010) (Fig. 6.3). The scale up process can possibly be made either with undifferentiated cells, at the definitive endoderm stage, or the liver progenitor stage. However, whatever method is chosen, the possibility to use a robotic system for expansion is desirable. In a move toward automation, feeder-free culture systems with defined media have recently been introduced, which greatly benefit the scale up process, and generate a clean system uncontaminated of animal components (Medine

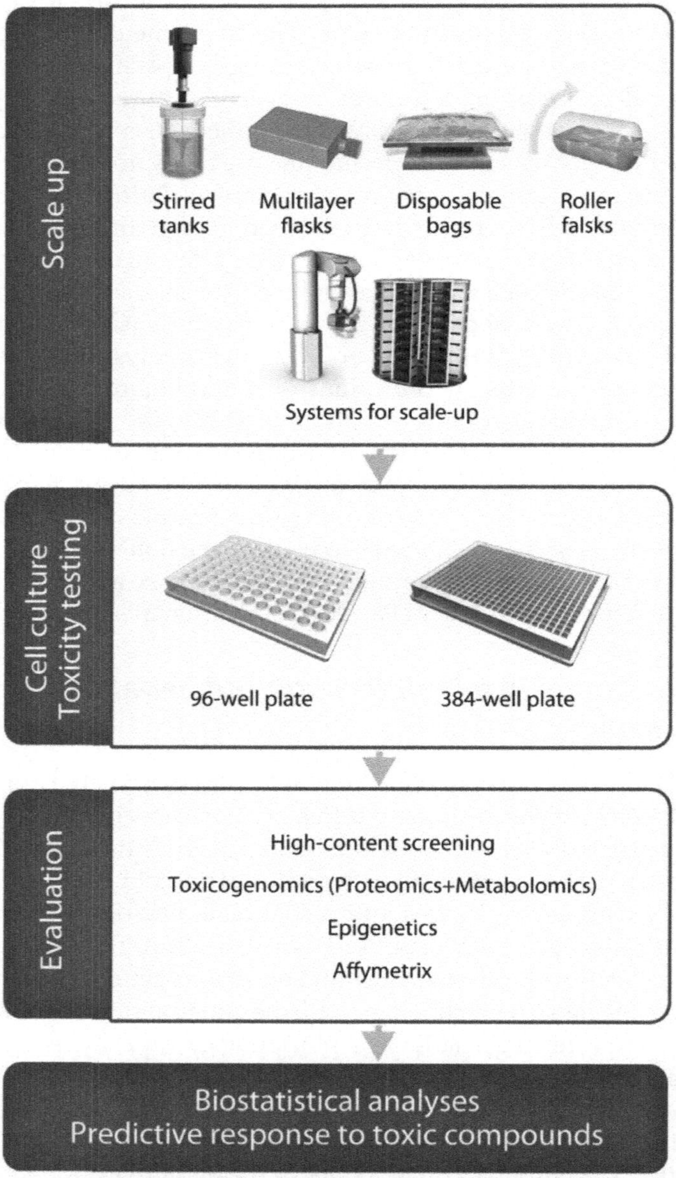

Figure 6.3. Illustration of steps essential for a high through-put hepatoxicity assay. First, the cell number should be possible to scale up, without loss of viability and pluripotency. Second, for cell cultures and toxicity testing in high through-put, the tests need to be performed in multiwell plates suitable for robot systems. Finally, to predict hepatocyte specific toxicity, evaluations with different methods and biostatistical analysis would benefit automated systems, such as high-content screening and affymetrix.

Color image of this figure appears in the color plate section at the end of the book.

et al. 2010, Vallier 2011). Well-defined controlled conditions are the key to successful large-scale long-term cultures, and as with the other steps in a predictive toxicity assay, the scale up process should be highly robust and reproducible (Medine et al. 2010). It is of relevance to keep this in mind when designing methods for high through-put *in vitro* toxicity methods.

Differentiation of hPSC to hepatocyte-like cells in a reproducible scaled up fashion is the next step in the development of successful hepatoxicity systems (Mandenius et al. 2011). As mentioned above, for drug development and safety assessment studies, it is desirable that the function of the cultured hepatocytes resemble the liver cells *in vivo* as far as possible. Therefore, great efforts are being made to develop large scale culture techniques allowing the cells to expand and differentiate. Several different approaches have been evaluated, both in 2D- and 3D format (Khetani and Bhatia 2008, Jensen et al. 2009, Upadhaya and Selvaganapathy 2009, Farzaneh et al. 2010, Mandenius et al. 2011, Miki et al. 2011, Zeilinger et al. 2011). For maintenance and differentiation of hESC derived hepatocytes, it is known that the microenvironment plays a central role (Underhill and Bhatia 2007). Cell-cell and cell-extra cellular matrix interactions, localized and gradient stimuli of soluble factors, and the 3D architecture are factors affecting the signalling pathways of survival and differentiation. For example, the 3D sandwich cultures with laminin, matrigel or collagen overlay have been shown to be beneficial for hepatic functionality (Ainscow et al. 2008). Although hepatocytes cultured in different 3D constructs have potential in toxicity testing, they add procedural complexity which is not conductive with high through-put screening. Ideally, the cells should be cultured as monolayer in multiwell plates, preferably 96- or 384-well plates. Furthermore, for pharmacological and toxicological studies, it is an advantage with methods and culture techniques that facilitate the differentiation and maintenance of mature phenotypes for prolonged culture periods. This is particularly important in view of the fact that exposure times should be long enough for some toxicologically relevant processes to occur. Identification of properties supporting differentiation and long-term maintenance, such as coating with the polyurethane matrix polymer 134, in multiwell plate formats is therefore highly interesting (Hay et al. 2011).

There are numerous methods for evaluating toxic effects of cells/tissues *in vitro* or *in vivo*, as described in depth in a review by Waters and Fostel (Waters and Fostel 2004). Comparison of data generated from hPSC-derived hepatocyte-like cells with other relevant cellular models generated *in vitro* has been proposed to help guide the risk assessment/decision-making process (Suzuki et al. 2010, Mandenius et al. 2011). As mentioned above, human primary hepatocytes are considered golden standard, and are therefore often used as a reference cell types in toxicity evaluations. However, we should bear in mind that human primary hepatocytes exhibit a great inter-experimental variability in both quality and metabolizing capacity, and hence, great care should be taken when selecting a reference cell system.

Combining several different assays with different endpoints is probably needed to predict toxicity (Gómez-Lechón et al. 2010). To that end, results from analysis of mitochondrial function, oxidative stress, accumulation of lipids and impaired bile acid transport could potentially be combined with morphological signs and a genomics approach. To increase the throughput of the techniques, different actions to speed up sample collection and analysis should be considered. An automated sampling process would be a great advantage for a high through-put toxicity testing. Depending on the type of evaluation method and read-out, this final step of the toxicity assay might be more complex to atomize. However, with a wise choice of read-out methods and criteria, the evaluation step could also be automated.

In order to make the hPSC-derived hepatocytes a strong *in vitro* hepatoxicity model, high fidelity and mature phenotype is required. We believe that further improvements of published systems are needed for an *in vitro* liver toxicity prediction model to be established, to be informative and therefore useful alternatives to current *in vitro* and *in vivo* methods. Once this has been achieved, these assays are likely to be recognized by regulatory agencies and the pharmaceutical industry. Hence, robust and reproducible hepatic differentiation protocols have to be developed before such system can be implemented into routine toxicity testing (Hengstler et al. 2005). For optimization of promising methods to meet industrial needs, it is essential to validate reproducibility and robustness of the technique. All steps of the experimental protocols, from cell quality and scale up to read outs, must be verified and controlled. If all the above mentioned steps are satisfied, we are on the way to building a robust and reliable predictive model which has potential to improve drug safety and reduce drug attrition.

Ongoing studies

Awareness of the current insufficient hepatoxicity assessment techniques, with limited access to human relevant model systems, has generated large investments for the development of alternative *in vitro* high through-put assays. Several large companies have started to establish new strategies based on hPSC for safety assessment (Baker 2010). In the UK, three big pharmaceutical companies together with UK authorities have invested significantly into the project Stem Cells For Safer Medicines (http://www.sc4sm.org/). The goal is to develop novel hepatotoxicity assays based on hPSC, but it remains to be seen if this effort will be successful, and will have freedom to operate in this space. An ongoing consortium, carcinoGENOMICS, founded by the European Commission was initiated 2006 with the aim to find alternatives to current rodent-based bioassays for genotoxicity and carcinogenicity (http://www.carcinogenomics.eu/). By applying a well-defined set of compounds known to cause such toxicity, the endeavour is to establish a robust and predictive *in vitro* system based on "omics" technologies. Indeed, from that project an *in vitro* model for predictive chemical carcinogenesis using hESC derived hepatocyte-like cells has recently been demonstrated by Yildirimman and co-workers (Yildirimman et al. 2011). In brief, by evaluating the toxic effect of 15 compounds, grouped into three categories based on known mechanism of toxicity, a remarkably small number of 592 predictive genes were found with Affymetrix microarrays. With this set of genes, it was possible to classify the three groups of test compounds. It was also shown that this genetic response strongly correlated with *in vitro* cultures of human primary hepatocytes. To our knowledge, this is the first study reporting a robust and predictive *in vitro* hESC-based hepatoxicity assay. This clearly illustrates the potential use of hESC for safety assessment, and the first step towards a predictive, human relevant, hepatotoxicity assessment assay. In another ongoing major international program named Stem Cell Research & Tox, a 5 yr collaborative research project co-funded by the European Commission and by COLIPA (the European Cosmetics Association), an additional hepatotoxicity investment has been made (http://www.scrtox.eu/). This project has been designed to make use of hPSC-derived hepatocytes to provide *in vitro* assays for predicting toxicity of pharmaceutical compounds and cosmetic ingredients, and to address all issues related with biological and technological resources to meet this goal. Furthermore, EU has currently an open call for research grant proposals named Innovative Medicines Initiative (IMI, http://www.imi.europa.eu/). This call focuses on early identification of drug induced toxicity, using a panel of newly developed assays, with special aim to predict DILI in humans. These research grants are co-founded by the European Commission and the pharmaceutical industry. In the above described projects, large investments are made from member states of the European Union, as well as from industry struggling with escalating costs and potentially hazardous products. Thus, the urgent need for alternative high through-put models predictive of hepatoxicity is clear.

Conclusion

Adverse reactions of approved drugs can have severe effects on the patients, and as a consequence, be of economic significance for the public healthcare. Furthermore, compound attrition and late stage drug withdrawal due to adverse effects is a major burden for the pharmaceutical industry. As the cost for development and introduction of a new drug to the market is thought to be several billion USD, the pharmaceutical industry is determined to increase the screening options at an early stage of product development.

Hepatotoxicity is traditionally assessed with animal experiments, cell cultures of primary hepatocytes from animals or humans, or immortalized hepatic cell lines. Unfortunately, all presently available techniques have their limitations, as toxicity tests of animal cells are poorly predictive of effects on humans and immortalized cell lines have, per definition, an altered cell cycle and death program. In addition, human primary hepatocytes are most often derived from clinical surgery of severely diseased patients and the variability between individuals is generally large. Thus, these obstacles drive the pharmaceutical and chemical industry to search for more reliable and robust models of hepatotoxicity.

Stem cells represent the most recent phase of the biotechnology revolution in medicine. Their significance is based on an understanding of how all cells develop from embryonic stem cells into adult tissues. Knowledge of these processes, together with the development of technologies for the manipulation of growth and differentiation will enable stem cells to be used as a potentially unlimited biological resource for cell-based human discovery tools for drug discovery and toxicity testing. However, several practical steps are important to evaluate and refine, in order to develop a predictive high through-put hepatoxicity assay. First, suitable scale up techniques for the starting material (hPSC) needs to be developed, with, e.g., bioreactors such as stirred tanks or multilayer flasks. Second, and perhaps the most important step, is a robust and reproducible differentiation into hepatocyte-like cells with protocols preferably applicable for multiple cell lines. The third step, is to evaluate the possibility to perform tests on numerous substances in an automated manner, ideally in multiwell plates. Fourth, the choice of methods of evaluation should be based on techniques that are amenable to high through-put, and finally, it is essential to perform thorough biostatistical analysis of the large amount of data generated by the evaluation methods. As proposed by several researchers, no single method would cover all possible aspects and alterations caused by toxic exposure. Thus, combined readouts from several different evaluation methods and cellular models will be required in order to build a predictive model.

Independent research groups have indeed presented endpoints of hepatic differentiation, maturation and effects of toxic exposure. Validated, standardized and joint consensus on the selection of biomarkers predictive of hepatocyte toxicity, presented for example in a database, would be useful for academic researchers, pharmaceutical industry as well as for regulatory agencies.

For a fast moving field of science, it is important to update both methodology and principles of evaluation. This is also the case for safety assessment studies. Surprisingly, the determination criteria for drug- safety testing has not changed in decades. There are currently several novel and potentially very informative and useful methods that has been established by both academic researchers and by the industry. However, for a test to be accepted as a safety testing method, both the FDA and the EMA require an independent validation. In 2006, FDA therefore initiated a new consortium, Predictive Safety Testing Consortium (PSTC), with the aim of bringing pharmaceutical companies, US and European regulatory agencies together to share knowledge to validate novel and improved safety testing methodology (http://www.c-path.org/pstc.cfm). This trans-Atlantic joint effort will provide knowledge and new methodology in safety assessment for drug discovery, and can be used by pharmaceutical scientists, regulators and clinicians. It is anticipated that new improved *in vitro* models based on physiologically relevant human cell systems will result in species-specific and more cost-effective assays, ultimately leading to safer new drugs and better evaluations of the effects of chemicals on human health.

In conclusion, as described in the present chapter, hPSC derived hepatocyte-like cells have been shown to be useful for more than clinical replacement therapies, and thus, we believe that possible future applications for hPSC face an infinite horizon. The first applications of hepatoxicity using hPSC-derived hepatoytes are in fact just around the corner.

References

Abboud G and N Kaplowitz. 2007. Drug-induced liver injury. Drug Safety. 30: 277–294.

Abraham VC, DL Towne, JF Waring et al. 2008. Application of a high-content multiparameter cytotoxicity assay to prioritize compounds based on toxicity potential in humans. J Biomol Screen. 13: 527–537.

Agarwal S, KL Holton and R Lanza. 2008. Efficient differentiation of functional hepatocytes from human embryonic stem cells. Stem Cells. 26: 1117–1127.

Ainscow EK, JE Pilling, NM Brown et al. 2008. Investigations into the liver effects of ximelagatran using high content screening of primary human hepatocyte cultures. Expert Opin Drug Safety. 7: 351–365.

Amacher DE 2010. The discovery and development of proteomic safety biomarkers for the detection of drug-induced liver toxicity. Toxicol Appl Pharmacol. 245: 134–142.

Améen C, R Strehl, P Björquist et al. 2008. Human embryonic stem cells: current technologies and emerging industrial applications. Crit Rev Oncol Hematol. 65: 54–80.

Asgari S, B Pournasr, GH Salekdeh et al. 2010. Induced pluripotent stem cells: a new era for hepatology. J Hepatol. 53: 738–751.

Baker M. 2010. Testing time for stem cells. Nature. 463: 719.

Barbaric I, PJ Gokhale, M Jones et al. 2010. Novel regulators of stem cell fates identified by a multivariate phenotype screen of small compounds on human embryonic stem cell colonies. Stem Cell Res. 5: 104–119.

Basma H, A Soto-Gutiérrez, GR Yannam et al. 2009. Differentiation and transplantation of human embryonic stem cell-derived hepatocytes. Gastroenterol. 136: 990–999.

Bickle M. 2010. The beautiful cell: high-content screening in drug discovery. Anal Bioanal Chem. 398: 219–226.

Brolén G, L Sivertsson, P Björquist et al. 2010. Hepatocyte-like cells derived from human embryonic stem cells specifically via definitive endoderm and a progenitor stage. J Biotechnol. 145: 284–294.

Cai J, Y Zhao, Y Liu et al. 2007. Directed differentiation of human embryonic stem cells into functional hepatic cells. Hepatology. 45: 1229–1239.

Duan Y, A Catana, Y Meng et al. 2007. Differentiation and enrichment of hepatocyte-like cells from human embryonic stem cells *in vitro* and *in vivo*. Stem Cells. 25: 3058–3068.

Duan Y, X Ma, W Zou et al. 2010. Differentiation and characterization of metabolically functioning hepatocytes from human embryonic stem cells. Stem Cells. 28: 674–686.

Ebert AD and CN Svendsen. 2010. Human stem cells and drug screening: opportunities and challenges. Nat Rev Drug Discov. 9: 367–372.

Farzaneh Z, B Pournasr, M Ebrahimi et al. 2010. Enhanced functions of human embryonic stem cell-derived hepatocyte-like cells on three-dimensional nanofibrillar surfaces. Stem Cell Rev. 6: 601–610.

Ghodsizadeh A, A Taei, M Totonchi, et al . 2010. Generation of liver disease-specific induced pluripotent stem cells along with efficient differentiation to functional hepatocyte-like cells. Stem Cell Rev. 6: 622–632.

Gómez-Lechón MJ, JV Castell and MT Donato. 2010. The use of hepatocytes to investigate drug toxicity. Methods Mol Biol. 640: 389–415.

Greenhough S, CN Medine and DC Hay. 2010. Pluripotent stem cell derived hepatocyte-like cells and their potential in toxicity screening. Toxicology. 278: 250–255.

Guguen-Guillouzo C and A Guillouzo. 2010. General review on *in vitro* hepatocyte models and their applications. Methods Mol Biol. 640: 1–40.

Hartung T and C Rovida. 2009. Chemical regulators have overreached. Nature 460: 1080–1081.

Hay DC, D Zhao, A Ross et al. 2007. Direct differentiation of human embryonic stem cells to hepatocyte-like cells exhibiting functional activities. Cloning Stem Cells. 9: 51–62.

Hay DC, J Fletcher, C Payne et al. 2008a. Highly efficient differentiation of hESCs to functional hepatic endoderm requires Activin A and Wnt3a signaling. Proc Natl Acad Sci USA. 105: 12301–12306.

Hay DC, D Zhao, J Fletcher et al. 2008b. Efficient differentiation of hepatocytes from human embryonic stem cells exhibiting markers recapitulating liver development *in vivo*. Stem Cells. 26: 894–902.

Hay DC, S Pernagallo, JJ Diaz-Mochon et al. 2011. Unbiased screening of polymer libraries to define novel substrates for functional hepatocytes with inducible drug metabolism. Stem Cell Res. 6: 92–102.

Hengstler JG, M Brulport, W Schormann et al. 2005. Generation of human hepatocytes by stem cell technology: definition of the hepatocyte. Expert Opin Drug Metab Toxicol. 1: 61–74.

Hewitt NJ, MJ Lechón, JB Houston et al. 2007. Primary hepatocytes: current understanding of the regulation of metabolic enzymes and transporter proteins, and pharmaceutical practice for the use of hepatocytes in metabolism, enzyme induction, transporter, clearance, and hepatotoxicity studies. Drug Metab Rev. 39: 159–234.

Hochedlinger K and K Plath. 2009. Epigenetic reprogramming and induced pluripotency. Development. 136: 509–523.

Hu Q, AM Friedrich, LV Johnson et al. 2010. Memory in induced pluripotent stem cells: reprogrammed human retinal-pigmented epithelial cells show tendency for spontaneous redifferentiation. Stem Cells. 28: 1981–1991.

Jensen J, J Hyllner and P Björquist. 2009. Human embryonic stem cell technologies and drug discovery. J Cell Physiol. 219: 513–519.

Kang HC. 2010. Disease-specific pluripotent stem cells. Korean J Pediatr. 53: 786–789.

Kang HC and JE Trosko. 2011. Stem cells in toxicology: fundamental biology and practical considerations. Toxicol Sci. 120: 269–289.

Kehoe DE, D Jing, LT Lock et al. 2010. Tzanakakis. Scalable stirred-suspension bioreactor culture of human pluripotent stem cells. Tissue Eng Part A. 16: 405–421.

Khetani SR and SN Bhatia. 2008. Nicroscale culture of human liver cells for drug development. Nat Biotechnol. 26: 120–126.

Kim K, A Doi, B Wen et al. 2010. Epigenetic memory in induced pluripotent stem cells. Nature 467: 285–290.

King JA and WM Miller. 2007. Bioreactor development for stem cell expansion and controlled differentiation. Curr Opin Chem Biol. 11: 394–8.

Kola I. 2008. The state of innovation in drug development. Clin Pharmacol Ther. 83: 227–230.

Lasher CD, P Rajagopalan and TM Murali. 2011. Discovering networks of perturbed biological processes in hepatocyte cultures. PLoS One. 6: e15247.

Laustriat D, J Gide and M Peschanski. 2010. Human pluripotent stem cells in drug discovery and predictive toxicology. Biochem. Soc Trans. 38: 1051–1057.

Liras A. 2010. Future research and therapeutic applications of human stem cells: general, regulatory, and bioethical aspects. J Transl Med. 8: 131.

Liu H, Z Ye, Y Kim et al. 2010. Generation of endoderm-derived human induced pluripotent stem cells from primary hepatocytes. Hepatology. 51: 1810–1819.

Lou YJ and XG Liang. 2011. Embryonic stem cell application in drug discovery. Acta Pharmacol Sin. 32: 152–159.

Mandenius CF, TB Andersson, PM Alves et al. 2011. Toward oreclinical predictive drug testing for metabolism and hepatoxicity by using *in vitro* models derived from human embryonic stem cells and human cell lines—a report on the Vitrocellomics EU-project. ATLA. 39: 1–25.

Medine CN, S Greenhough and DC Hay. 2010. Role of stem-cell-derived hepatic endoderm in human drug discovery. Biochem Soc Trans. 38: 1033–1036.

Meissner A. 2010. Epigenetic modifications in pluripotent and differentiated cells. Nat Biotechnol. 28: 1079–1088.

Miki T, A Ring and J Gerlach. 2011. Hepatic Differentiation of Human Embryonic Stem Cells Is Promoted by Three-Dimensional Dynamic Perfusion Culture Conditions. Tissue Eng. Part C Methods. Feb 15. [Epub ahead of print]

O'Brien PJ, W Irwin, D Diaz et al. 2006. High concordance of drug-induced human hepatotoxicity with *in vitro* cytotoxicity measured in a novel cell-based model using high content screening. Arch Toxicol. 80: 580–604.

Olson H, G Betton, J Stritar et al. 1998. The predictivity of the toxicity of pharmaceuticals in humans from animal data—an interim assessment. Toxico Lett. 102–103: 535–538.

Olson H, G Betton, D Robinson et al. 2000. Concordance of the toxicity of pharmaceuticals in humans and in animals. Regul Toxicol Pharmacol. 32: 56–67.

Phillips BW and JM Crook. 2010. Pluripotent human stem cells. A novel tool in drug discovery. Biodrugs 24: 99–108.

Polo JM, S Liu, ME Figueroa et al. 2010. Cell type of origin influences the molecular and functional properties of mouse induced pluripotent stem cells. Nat Biotechnol. 28: 848–855.

Rashid ST, S Corbineau, N Hannan et al. 2010. Modeling inherited metabolic disorders of the liver using human induced pluripotent stem cells. J Clin Invest. 120: 3127–3136.

Rovida C and T Hartung. 2009. Re-evaluation of animal numbers and costs for *in vivo* tests to accomplish REACH legislation requirements for chemicals—a report by the transatlantic think tank for toxicology (t(4)). ALTEX. 26(3): 187–208.

Sancho-Bru P, M Najimi, M Caruso et al. 2009. Stem and progenitor cells for liver repopulation: can we standardise the process from bench to bedside? Gut. 58: 594–603.

Sartipy P and P Björquist. 2011. Concise review: Human pluripotent stem cell-based models for cardiac and hepatic toxicity assessment. Stem Cells. 29: 744–748.

Sartipy P, P Björquist, R Strehl et al. 2007. The application of human embryonic stem cell technologies to drug discovery. Drug Discov Today. 12: 688–699.

Serra M, C Brito, MF Sousa et al. 2010. Improving expansion of pluripotent human embryonic stem cells in perfused bioreactors through oxygen control. J Biotechnol. 148: 208–215.

Shi Q, H Hong, J Senior et al. 2010. Biomarkers for drug-induced liver injury. Expert Rev Gastroenterol. Hepatol. 4: 225–234.

Shiraki N, K Umeda, N Sakashita et al. 2008. Differentiation of mouse and human embryonic stem cells into hepatic lineages. Genes Cells. 13: 731–46.

Si-Tayeb K, FP Lemaigre and SA Duncan. 2010a. Organogenesis and development of the liver. Dev Cell. 18: 175–189.

Si-Tayeb K, FK Noto, M Nagaoka et al. 2010b. Highly efficient generation of human hepatocyte-like cells from induced pluripotent stem cells. Hepatology. 51: 297–305.

Sullivan GJ, DC Hay, IH Park et al. 2010. Generation of functional human hepatic endoderm from human induced pluripotent stem cells. Hepatology. 51: 329–335.

Suzuki A, RJ Andrade, E Bjornsson et al. 2010. Drugs associated with hepatotoxicity and their reporting frequency of liver adverse events in VigiBase: unified list based on international collaborative work. Drug Saf. 33: 503–522.

Takahashi K and S Yamanaka. 2006. Induction of pluripotent stem cells from mouse embryonic and adult fibroblast cultures by defined factors. Cell. 126: 663–676.

Touboul T, NR Hannan, S Corbineau et al. 2010. Generation of functional hepatocytes from human embryonic stem cells under chemically defined conditions that recapitulate liver development. Hepatology. 51: 1754–1765.

Trosko JE and CC Chang. 2010a. Factors to consider in the use of stem cells for pharmaceutic drug development and for chemical safety assessment. Toxicology. 270: 18–34.

Trosko JE and BL Upham. 2010b. A paradigm shift is required for the risk assessment of potential human health after exposure to low level chemical exposures: a response to the toxicity testing in the 21st century report. Int J Toxicol. 29: 344–357.

Underhill GH and SN Bhatia. 2007. High-throughput analysis of signals regulating stem cell fate and function. Curr Opin Chem Biol. 11: 357–366.

Upadhaya S and PR Selvaganapathy. 2009. Miniaturized microfluidic formats for cell-based high-throughput screening. Crit Rev Biomed Eng. 37: 193–257.

Vallier L. 2011. Serum-free and feeder-free culture conditions for human embryonic stem cells. Methods Mol Biol. 690: 57–66.

Waters MD and JM Fostel. 2004. Toxicogenomics and systems toxicology: aims and prospects. Nat Rev Genet. 5: 936–948.

Wobus AM and P Löser. 2011. Present state and future perspectives of using pluripotent stem cells in toxicology research. Arch Toxicol. 85: 79–117.

Wu MH, SB Huang and GB Lee. 2010. Microfluidic cell culture systems for drug research. Lab Chip. 10: 939–956.

Yildirimman R, G Brolén, M Vilardell et al. 2011. Human embryonic stem cell derived hepatocyte-like cells as a tool for *in vitro* hazard assessment of chemical carcinogenicity. Toxicol Sci. 124: 278–90.

Zeilinger K, T Schreiter, M Darnell et al. 2011. Scaling Down of a Clinical Three-Dimensional Perfusion Multicompartment Hollow Fiber Liver Bioreactor Developed for Extracorporeal Liver Support to an Analytical Scale Device Useful for Hepatic Pharmacological *in vitro* Studies. Tissue Eng. Part C Methods. Feb 16. [Epub ahead of print]

7

Stem Cells in the Development of Products for Regenerative Medicine

Paul A. De Sousa,[1,a,*] *Scott McRae*[2] and *Glyn Stacey*[3]

Introduction—the challenge of culturing pluripotent stem cells

In research, the isolation and cultivation of any new type of cell begins with a dependence on the supportive environment provided by crude factors such as helper cells, serum and other bio-active factors. The administration of these factors is equally dependent on an insightful operator attuned to cellular responses and in particular the appearance of the cells. Progression beyond this artisan-like beginning is founded on fundamental and empirical research into the intrinsic and extrinsic factors which define cellular phenotype, upon which supportive defined media and matrices are ultimately developed.

Thanks to decades of industrial bioprocessing for the production of cellular by-products such as antibodies, vaccines, and bio-active proteins there exists today a broad range of semi- or fully automated bioreactor

[1]Scottish Centre for Regenerative Medicine, University of Edinburgh, Edinburgh, EH16 4SB, UK.
[a]Email: paul.desousa@ed.ac.uk
[2]Cell Guidance Systems, Babraham Research and Control-HPA Campus, Cambridge, UK; Email: scott.mcrae@cellgs.com
[3]National Institute for Biological Standards-HPA, Blanche Lane, South Mimms, Herts, EN6 3QG. Email: glyn.stacey@nibsc.hpa.org.uk
*Corresponding author

technologies for cellular expansion, some of which have the capacity to provide continuous monitoring of physical parameters of the culture environment as indirect indices of cell viability. These include measurement of pH, temperature and nutrient consumption. An optimal and universal system for stem cell culture does not exist, although a spectrum of systems for adherent and non-adherent cell processing have been applied successfully in specific contexts (reviewed by Placzek et al. 2009). Obviously, the utility of these systems is based on their ability to stably preserve the distinctive properties of the target cell being cultivated. Irrespective of their source or nature, the most distinctive property of stem cells which sets them apart from differentiated or immortalized cell lines is their dynamic and dual capacity for growth or to be specified into other cells. These fates are understood to be determined by both intrinsic (i.e., gene expression, metabolism, nuclear and cyto-architecture) and extrinsic factors (i.e., bio-active factors in their environment such as extracellular matrix, soluble growth factors and organic and inorganic molecules), the complexities of which are context specific and still being unravelled. Both growth and differentiation potential vary between stem cell types and it is this variation which normally defines the specific challenges to cultivate them *in vitro*. It follows that the greater the potential to grow or specify other cells, the more difficult it is to maintain them *in vitro* as a homogeneous population. Given that cells in culture will compete with each other for nutrients and exogenously supplied bio-active factors, and they will stimulate each other in a paracrine fashion, cellular heterogeneity resulting from spontaneous or uncontrolled differentiation can effectively destabilize a culture environment and its efficacy.

Whether isolated from pre-implantation embryos or induced by genetic modification of adult cells (Thomson et al. 1998, Takahashi and Yamanaka 2006), pluripotent stem cells have the greatest capacities for growth and differentiation and it is these properties that have substantiated enthusiasm for their use in research, industrial and therapeutic applications. However, the breadth of these capacities also makes them more difficult to cultivate efficaciously than cells with single or more restricted lineage potency. The fundamental challenge in the manufacture of pluripotent stem cells for regenerative medicine is thus centred on the control and qualification of their dynamic nature. This begins with defining the *in vitro* systems which sustain them, the utility of which is ultimately determined by their capacity to be consistently and applied across cell lines irrespective of their origin. In this chapter we begin with a historical perspective on the roots of culture systems supportive of human pluripotent stem cells as originally isolated from early embryos through to the evolution of defined culture systems now in practice. We discuss complimentary aspects of cryopreservation and scalable expansion followed by the challenge of product qualification. Lastly we discuss the regulatory requirements to manufacture pluripotent stem cells for therapeutic use, with a focus on the European and UK experience. Our objective is to frame where we have been, are and have yet to get to so that we can realize the translational promise of such a unique and powerful resource for regenerative medicine.

Definition of culture systems supporting hESC renewal

Similar to early studies in the mouse, the first culture system to maintain hESC renewal *in vitro* relied heavily on crude animal sourced biological reagents. This included mitotically inactivated mouse embryo fibroblasts (MEFs) as a feeder cell substrate for attachment or to precondition media, coupled with media supplementation with bovine fetal serum (Thomson et al. 1998). Since then there has been a gradual improvement in the definition of media environments, beginning with the provision of first animal and then human crude or purified biologics such as serum fractions, commonly referred to as serum replacements, through to the provision of recombinant growth factors, (reviewed by Rao and Zandstra 2005). The first feeder-free culture system supporting hESC renewal, reported by Xu et al. (2001), involved culture on a substrate of Matrigel™, a crude poorly defined extracellular matrix preparation sourced from a murine sarcoma (EHS). This required preconditioning of media by exposure to mitotically inactivated MEFs for 24 hr and further supplementation with basic Fibroblast Growth Factor (bFGF) and KNOCKOUT™ serum replacement (KOSR), a proprietary formulation containing lipid enriched bovine serum albumin. In the quest to make xeno-free media formulations independent of KOSR, alternative culture systems reliant on other proprietary products of unknown composition have been developed (for example see Yao et al. 2006). While from a commercial perspective, reliance on proprietary products of unknown formulation is understandable, ignorance of a media system's composition is problematic when it is used for research or when the system fails in a manufacturing context. Matrigel™ can be replaced by purified extracellular matrix proteins such as human laminin, fibronectin and vitronectin, presented either alone or as a blend. These substrates can support hESC isolation and renewal in different chemically defined media supplemented with bFGF and a range of other growth factors (Li et al. 2005, Fletcher et al. 2006, Ludwig et al. 2006). Although this has represented a significant advance in the field, reliance on purified biologics is still challenged by both variations in the bio-activity and current reliance on frequently animal sourced reagents for their preparation. An example of this is the fact that isolation and culture of human fibroblast feeders, which can substitute for MEFs as media preconditioners, still involves exposure to Low Serum Growth Supplement (LSGS) containing 2% (v/v) fetal bovine serum. Although this serum is sourced from cattle certified to be free of Transmissible Spongiform Encephalopathy (TSE) in New Zealand, the potential for introduction of as yet unknown pathogens can never be ruled out. In fact most sources of bovine serum are known to be contaminated with bovine viral diarrhoea virus (BVDV) and commercial sources of bovine serum have been found to contain more than one bovine virus. This is relevant to procedures used in the purification or recombinant production of human proteins such as albumin, extracellular matrix proteins and growth factors, which may rely on animal sourced reagents.

Human ESCs differ from mESCs in the signalling pathways which sustain them. For some time this has been simply characterized by a requirement of the former for bFGF, compared with LIF in the mouse. However, this is likely to be an oversimplification of complex signalling pathways. For example, TGFβ1, Leukemia Inhibitory Factor (LIF), and the Bone Morphogenetic Protein (BMP) antagonist Noggin have all been suggested to synergize with bFGF to suppress differentiation and sustain self-renewal of hESCs (Amit et al. 2000, Xu et al. 2005). Other reports have described chemically defined culture systems supporting self-renewal containing bFGF together with Wnt3A and a B-cell-activating factor (Lu et al. 2006). Dependence on bFGF has also been described as a function of paracrine signalling between undifferentiated and differentiating cells in culture. Specifically, hESC renewal has been proposed to be mediated by IGF-II and TGF-β secreted from spontaneously differentiating fibroblast-like cells which are the targets of bFGF (Bendall et al. 2007). Irrespective of whether this is in fact the case or only a reflection of the feeder-conditioned media system in which these experiments were conducted, the importance of paracrine signalling within *in vitro* cellular microenviroments cannot be underestimated. Empirically the enigmatic relationship between hESC can be manipulated to maintain undifferentiated hESC renewal such as for example following repeated passaging of hESCs dissociated to single cells. This normally will prompt hESC differentiation unless cells are maintained at high density, presumably as a result of facilitated paracrine stimulation. Potency of the hESC microenviroment to regulate cell phenotype has been linked to epigenetic regulation of the TGFβ family (Hendrix et al. 2007). The importance of TGFβ signalling as a means of maintaining human pluripotent stem cells centres around co-activation of the FGFR and TGFβ pathways to control the stability of SMAD4, either through activation of the TGFβ receptors, or inhibition of BMP activity (Xiao et al. 2006, Avery et al. 2010). This is thought to be in concert with additional tyrosine kinase receptor ligands such as IGFR or EGFR agonists (Wang et al. 2007).

Although much progress has been made towards feeder- and xeno-free culture conditions (Ludwig et al. 2006, Lu et al. 2006, Yao et al. 2006), the conditions required are still complex and reliant on human derived reagents which can be subject to a lot of variation and contamination with human pathogens. One means by which researchers have sought to obviate reliance on biological components supporting stem cell renewal or induction of differentiation has been through the use of cell permeable "small molecule" chemicals for high affinity targetted stimulation/interference with intracellular signalling cascades driving these processes. In hESCs this has been exemplified through pharmacological interference with Wnt and RasGAP/ERK pathways (Sato et al. 2004, Chen et al. 2006). However, the holy grail of small molecule maintenance of the hESC phenotype has not been a simple quest. Numerous high content/moderate throughput small molecule screening programs, published and unpublished, have failed to yield the single "magic bullet" capable of sustaining pluripotency (Desbordes et al. 2008, Andrews et al. 2010, Damoiseaux et al. 2009). Some molecules were claimed to possess such efficacy retain dependency on conditioned medium, failing to find application in a defined environment (Burton et al. 2010). The pre-eminent system in use today relies on small molecule activation of the Wnt and GABA receptors alongside bFGF and TGFβ1 (Ludwig et al. 2006). More recent developments include the use of combinations of kinase inhibitors targetting Rho kinase, glycogen synthetase kinase, FGFR tyrosine kinase and mitogen activated protein kinase pathways in conjunction with bFGF (Tsutsui et al. 2011). One common thread between each of these systems is the activation of Wnt signalling, either through GSK inhibition or by Wnt mediated activation and subsequent promotion of CBP/β-catenin association at the expense of the p300/β-catenin interaction. A mechanism which interestingly appears conserved between hESC and mESC (Miyabayashi et al. 2007). hESCs themselves are an *in vitro* artefact, derived from a cell population which is poorly defined and transitory. Studies of mouse ESCs has led to the proposal of a "ground state", in which blockage of the differentiation pathways is sufficient to commit the cells to self renew. This has lead to speculation that similar approaches may be feasible in hESC, and this has lent credence by the use of small molecules as described above. Intriguingly it has been demonstrated that it is feasible to sustain hESC for prolonged culture under conditions known to stabilize mESCs, utilizing MEK and p38 inhibitors (Xu et al. 2010). In this situation the hESC adopted a mESC related phenotype, indicating that self renewal of distinctly different states of pluripotent human cells can be achieved by different culture systems.

The issues of variability between culture systems are compounded by issues of variability between cell lines. How universally applicable variations in the required culture environment are across hESC lines is a question which is only now starting to be addressed through international coordination of laboratories such as, for example, the International Stem Cell Initiative (ISCI 2010). This is a question which has garnered enhanced significance as the questions of variability between hESC lines are matched equally by concerns over the uniformity of phenotype within populations. The potential for variability is evidenced by the differential contribution of clonal derivatives to different developmental end points (King et al. 2009, Stewart et al. 2010). The likelihood for subsets within or between self renewing populations of hESCs to contribute to lineage-specific end points emphasizes the need for fine control within systems destined for therapeutic or industrial application, an impossible goal when culture systems are based on undefined components, and the major driver towards the ultimate aim of fully synthetic culture.

The influence of the culture system can extend far beyond simply variation in the pluripotent state and lineage preference during commitment to differentiation. Indeed much of the analysis to date on the performance of culture systems over extended passaging has focused on the phenotypic stability of the population and its subsequent ability to follow particular differentiation protocols. However, it has been well established that hESC, and induced pluripotent stem cells iPSC, are prone to the acquisition of genomic abnormalities in culture frequently associated with malignant transformation (Maitra et al. 2005, Ben-David et al. 2010). The validation of novel culture systems for genomic stability is usually carried out by karyotyping, insufficient for the detection subchromosmal genetic change which has been shown to be associated with the long term passage of human pluripotent cells (Laurent et al. 2011) raising the question of whether this is a result of the selection pressures of cell culture, or reflects the inherent propensity of pluripotent cells in culture to be genetically unstable. Holubcová et al. (2011) have described a potential mechanism for this through centrosomal overamplification and demonstrate that improving binding to the culture substratum through application of fibronectin and laminin decreased the frequency of mitoses with supernumerary centrosomes. This gives a direct link between culture condition and genomic dysregulation, lending weight to the argument that such changes are induced by the culture environment and are not simply stochastic events sustained by the culture system. Further evidence of the ability of culture systems to mediate changes in the fate potential of stem cells has been demonstrated in numerous reports, proving that choice of matrix, presence of feeder cells, passaging method and hypoxic conditions can all influence either the transcriptome or epigenome of pluripotent cells (Allegrucci et al. 2007, Forsyth et al. 2008, Roobrouck et al. 2011). This influence of culture environment leaves the field in some difficulty since, increasingly, commercially available culture formulations are undisclosed "black boxes" making it extremely difficult for researchers to define appropriate, or even standard conditions for pluripotent cell line generation and expansion. Such systems are of limited value to the field since the lack of clarity in their composition precludes further advances in the state of the art of knowledge on how hESC renewal is controlled.

In summary, overcoming the challenges of inadvertent microbial contamination and persistence, variations in bioactivity of growth media, and maintaining genomic stability ultimately necessitate the development of biochemically equivalent substitutive technology to obviate current reliance on animal and human reagents of biological origin both for cell culture and storage. In developing this technology for safe clinical and efficient industrial application the need for traceability and transparency becomes ever more apparent.

Automated systems for pluripotent stem cell expansion

Ultimately the exploitation of pluripotent stem cell lines in screening systems for drug development or toxicology and for therapeutic application will require their scaled production either manually or using semi- or fully-automated systems. In any case, production systems must meet rigorous quality standards to ensure uniformity of product. Initial efforts seeking to mechanize aspects of manual picking of pluripotent stem cell colonies were impaired by the need to be able to adapt the culture technique to the cell condition. Extended passage of hESC using a flask-based culture in a fully automated system, for example in the form of the CompacT Select™ (TAP Biosystems, Royston, UK) has been demonstrated, both with conditioned medium (Thomas et al. 2009) and under chemically defined conditions (Cellartis AB, unpublished data). This technology can be further enhanced by inhibiting the apoptotic response to single cell dissociation of hESC using compounds capable of Rho kinase inhibition to achieve high efficiency replating (Andrews et al. 2010). The combination of Rho kinase inhibition with automation, enhanced the feasibility of using chemically defined culture systems in such processes. Combining the same process with Cell Factories could enable significant scale up, however, inter-passage variation in OCT4 expression and cyclical variation of proliferation rates have been reported (Thomas et al. 2009) suggesting that such methods will still fall below the required industry standard. The ongoing evolutionary aim of this process is to improve stability of phenotype alongside further gain in scalability, by progressing from planar or 2-dimensional (2-D) culture to 3-dimensional (3-D) culture system formats. These effectively provide the opportunity to cultivate more cells per unit of medium over the same period of time (Davis 2007). They also provide advantages over typical static culture systems in providing a more homogeneous culture environment where conditions such as temperature, pH and oxygen concentration can be monitored and controlled. Following from prior studies on mouse embryonic stem cells, a limited of number of studies have demonstrated feeder-free 3-D expansion of pluripotent hESC by a range of microcarrier dependent and independent means. The former include a range of microbeads (Phillips et al. 2008, Oh et al. 2009) or hydrogels differing in composition and porousity, providing either or both surface substrates for cell attachment or encapsulation. Reversible encapsulation of cells by active polymerization/depolymerization of a monomer subunit is the key feature of hydrogels, the efficacy of which for hESC expansion has been exemplified using hydrogels composed of polyacrylates (Li et al. 2006), Hyaluronic acid (Gerecht et al. 2007) and alginic acid and gelatin (Siti-Ismail et al. 2008). HESC expansion can also be achieved in the absence of carriers using stirred suspension bioreactors, through controlled application of shearing force and manipulation of signalling pathways mediating control of cell adhesion and cell death (Krawetz et al. 2009). However, there has been no direct comparison of these means and reported success may be dependent on the cell line used in the study or other soluble media factors, hydrogel or carrier free methods may offer the best expansion capabilities, whilst preserving pluripotent cell properties, with reports of up to 30 fold vs 10 fold expansion in 2D culture over the same culture period.

Cryopreservation of stem cells

Cryopreservation is an absolute requirement for long-term banking of cell cultures and may require careful development for new culture types, as in the case of hESCs. Methods for the cryopreservation of hESC have been based on empirical application of methods of conventional freezing or vitrification (Reubinoff et al. 2001, Fujioka et al. 2004, Suemori et al. 2006) developed for other similar cell types such as the human blastocyst (Vanderzwalmen et al. 2003) and mouse embryonic stem cells (mESCs). Conventional cryopreservation, whilst widely effective for continuous cell lines, has met with notable failures for hESCs (Heng et al. 2005), although more recent studies have indicated that high survival is obtainable with conventional slow cooling preservation techniques (for a recent review see Hunt 2011). Vitrification is the solidification of a liquid by increasing cryoprotectant concentration to high levels, inducing a glass structure; without crystallization and without the formation and growth of ice. The vitrification method, while capable of producing relatively good recovery is technically challenging due to the requirement to make serial transfers of cells from one cryoprotectant solution to another and finally to achieve super rapid cooling. The narrow bore "straws" typically used to enable the ultra-rapid cooling are also prone to contamination during processing and in liquid nitrogen storage. Despite the practical difficulties and risk of contamination, associated with vitrification it has been the method of choice for cyropreservation in cell banks. While this method can be used to bank 100–200 straws per cell line, the process is labour-intensive, involves complex transfers of cells, is operator dependent and prone to variation.

Cryopreservation media are usually composed of permeating and/or non-permeating cryoprotectants (CPs), which elicit cryoprotection through a complex process including their influence on osmosis or in some cases on the viscosity of the medium VEs? Combination of cryoprotective molecules of various classes allows a synergistic effect and a reduction of their individual concentration and toxicity. Classically, both serum and serum proteins have been used to enhance survival of cells. There are also natural anti-freeze (glyco-) proteins (AF(g)Ps) that prevent ice formation that may prove valuable in preservation technology. It has been demonstrated that some synthetic or non-mammalian AF(g)Ps can advantageously substitute for those of vertebrate origin (Stojkovic et al. 2002, Inglis et al. 2006, Lane et al. 2003). Theoretically, in the case of the latter the evolutionary distance of such species from humans could help reduce the risk of cross-specific transmission of pathogens. However, the possibility of new potential contaminants from reagents from unusual sources such as marine invertebrates and plants should be born in mind in evaluation of risk.

In summary, at the present time there is still scope for improvement in the cryopreservation of human pluripotent stem cells for industrial or clinical application. These improvements centre on the need to ensure the highest yields of live cell recovery where the differentiation state as well as morphological, cytogenetic and genomic status remain unchanged. As with human pluripotent stem cell culture, improvements will require replacement of undefined biological components with defined molecules and a shift from reliance on skilled operators to automatable systems.

Specification of the pluripotent cells as starting materials for cellular products

As described above, culture systems capable of hESC maintenance have in principal been in place for many years, but in practice there is still a requirement for improvement in this field. Our evolving knowledge of signalling pathways mediating the renewal of hESCs provides new opportunities to establish meaningful criteria to standardize the process of pluripotent cell generation and expansion. We can utilize a range of biological and molecular tools to monitor the status of the cellular phenotype through transcriptional and epigenetic profiling. This will be important in establishing parameters to which future cell products must conform. Traditionally hESC are evaluated according to a panel of accepted gene expression markers, at both protein and mRNA level, assessment of their capacity to differentiate and confirmation of a euploid karyotype. In addition to biochemical and molecular characterization of the cells, physical analysis can contribute to assessing the viability and differentiation status of hESC cultures. The ultrastructral characteristics of hESCs retain similarities to the early embryo ICM (Park et al. 2004, Cho et al. 2006) with marked changes at the onset of differentiation (Mumaw et al. 2010, Oh et al. 2005, Sathananthan and Trounson 2005). The organelle and ultrastructural characteristics of hESC include abundant ribosomes and polyribsomes, immature round mitochondria, large nuclei to cytoplasmic ratios and extensive microvilli, and are subject to substantial change as the cells commit to differentiation. More recently, a relatively specific molecular signature has been attributed to human embryonic stem cells and with an improved means of identification has come a better understanding of their potential applications. Among the surface markers used for immunoflouresence are glycolipids and glycoproteins including SSEA3 and 4, and TRA1-60 and 1–81, these in combination with the mRNA and protein expression of the key transcription factors regulating stemness, OCT3/4, SOX2, and NANOG, give solid indication of the pluripotent state and will undergo downregulation on commitment to differentiation. However, these markers are not sufficient in themselves to characterize subpopulations of hESC, and although they will ultimately silence, they are not sufficiently dynamic to detect early onset differentiation events. Gene expression profiling has been carried out extensively to analyze the hESC transcriptome, by microarray (ISCI 2007), SAGE (Richards et al. 2004) and massively parallel signature sequencing (Närvä et al. 2010) such analysis revealed that in many respects hESC maintain similarities to embryonal carcinoma cells but also reveal the core transcriptional networks controlling pluripotency. The development of array based QPCR technique such as Taqman Low Density Array (TLDA) allows us to exploit such databases by measuring quantitatively the expression of key groups of genes within the population. In this respect it is critical to establish both the presence of traditional accepted pluripotency markers as well as lack of expression of genes associated with commitment to differentiation, confirming the absence of uncontrolled differentiation events. In principle, such technology allows us to analyze the molecular signature of the population at critical points during the culture process.

There is an important delineation to be made between the characterization of a population of cells as opposed to the characterization of the cells within that population. When working with dynamically unstable populations of cells whose individual phenotypes are subject to variation in response to a continuously changing microenvironement it is possible that an apparently static global equilibrium can mask a very real fluidity in the population dynamic. Such fluidity means that at any one time we can expect variation in the pluripotent state and the readiness to differentiate individual cells. This must be factored in to the development of maintenance and differentiation protocols, but equally must be accounted for in the specification of the pluripotent phenotype. In order to account for this we have to look at means of characterizing these subpopulations in a quantifiable manner, this can only be achieved by expression analysis at the level of the single cell. Advances in analytical techniques in both QPCR and flow cytometry enable multi-parameter analysis of single cell gene expression, techniques which have already been applied in identifying sub-populations of embryonic stem cell-like cells from bone marrow (Sovalat et al. 2011) and could be readily adapted to the same purpose for hESC. Reports of differential expression of IGF1R and CD133 indicate that such subpopulations exist and are known to have varying intrinsic preferences in differentiation commitment (Stewart et al. 2010, King et al. 2009).

In addition to expression of marker genes it is evident that genomic stability is a prerequisite of any pluripotent culture and high priority needs to be placed on the absence of genetic changes which may potentiate neoplastic transformation. Aneuploidies are easily detected by karyotype but smaller abnormalities have been detected through comparative genomic hybridization (CGH) and single nucleotide polymorphism genotyping (SNP) (Lefort et al. 2008, Närvä et al. 2010, Spits et al. 2008, Laurent et al. 2011). High resolution SNP has identified a number of abnormalities specifically associated with the *in vitro* expansion of pluripotent cells which are otherwise undetectable by less sensitive methods, 90% of duplications in the genome are under 10 megabases in length (Laurent et al. 2011). In addition the same study reported that some regions appeared prone to the acquisition of genomic aberration in multiple unrelated pluripotent cell lines, raising concern regarding oncogenic potential of hESC. Indeed some of these regions are in proximity to genes known to regulate pluripotency. Some hESC lines have been shown to harbour features of neoplastic progression, but fail to demonstrate metastatic potential *in vivo* (Werbowetski-Ogilvie et al. 2009). It was suggested in this case that the transformed cells differentiated themselves from normal hESC by coexpression of FGFR1 and IGF1R, as opposed to only IGF1R. Importantly in this case it was shown that the differentiated progeny retained the genomic abnormalities of the parent cells, indicating that transformed hESC were not selected out of culture during differentiation. This confirms that functional assays are required to distinguish normal from pre-neoplastic cells and whilst the transformation potential of many chromosomal abnormalities remain unproven and their potential for use as markers of oncogenicity is limited.

Uncontrolled epigentic phenomena can also be important contributors to post-derivation effects on hESC stability, and may influence their response to differentiation stimuli (Tanasijevic et al. 2009). Expression of XIST and X chromosome inactivation (XCI) status can vary widely in different cultures of the same line, and different colonies within the same culture can be mosaic for XCI states. Indeed DNA methylation profiles have been shown to drift with extended passage of numerous hESC lines (Bibikova et al. 2006). One problem with determining the relevance of this potential for drift is that the likely implications for clinical application are not clearly understood. In specifying "normality" as a requirement for therapy directed hESC populations, we have to identify the current limitations in recognizing, detecting and interpreting genomic and epigenetic aberration. Technical difficulties need to be addressed in the high resolution analysis required for small genomic defects, particularly if they are present as low grade mosaicism, and in determining whether they are likely to manifest in a clinically significant manner. Of some comfort is the knowledge that the hESC microenvironment has been shown to inhibit or even reverse aggressive cancerous phenotypes (Postovit et al. 2008, Costa et al. 2009), and that under extended manual passage on human feeders hESC's have been demonstrated to maintain genomic stability even using high resolution techniques (Catalina et al. 2009).

Regulation of pluripotent stem cell based therapies—from a UK/EU perspective

The prospective governance of human pluripotent stem cell derived therapies in the future will likely vary around the world in accordance with the jurisdiction of the nation they are developed in. This will undoubtedly be influenced by a spectrum of cultural factors, beginning in the first instance on whether there is consensus on the usage of human embryos for non-reproductive purposes. To date there is still no internationally unified framework governing standards of quality and safety for cell and tissue therapies, although the recognition of a need for one has existed for some time, notably by the World Health Organisation (WHO) in 2003 in their report on access and safety in tissue and organ (see: WHO/HTP/EHT/T-2003.1; online @ http://www.who.int/ethics/topics/en/madrid_report_final.pdf). However, a significant milestone in the harmonization of European standards was reached with adoption by European Parliament in April 2004 of the EU Tissue and Cells Directive (2004/23/EC), with associated technical annexes published recently (2006/17/EC & 2006/86/EC). The EUTCD was based on standards previously established for autologous and allogeneic transplantation of blood tissues and organs by the Joint Accreditation Committee for the International Society for Cellular Therapy (ISCT) and the European Group for Blood and Marrow Transplantation (EBMT)- also know as JACIE; (http://www.jacie.org). The technical annexes were drawn largely from the UK Tissue Banking Code of Practice (2001) and its principles of providing a less prescriptive and brief guidance document. Collectively these set new minimum standards of quality and safety for the donation, procurement, testing, processing, storage, and distribution of human tissue and cells intended for human application. It broadly applies to cells derived from blood, bone marrow, reproductive cells (i.e., eggs, sperm, and embryos), foetal and adult tissues, and embryo stem cells. However, this directive excludes autologous grafts of tissues or cells within the same surgical procedure, allogeneic transplantation of blood, blood products, and tissues or cells or animal origin for which earlier directives and recommendations exist (i.e., Directives 2001/83/EC & 2000/70/EC, 2002/98/EC and Recommendation 98/463/EC). Also excluded are the uses of human tissues for research unless those tissues are used in clinical trials applied to the human body (De Sousa et al. 2006).

Further harmonization of standards with the EU was achieved by the European Parliaments' approval of the Advanced Therapies Directive for Medicinal Products (2005/0227/EC) in April 2007, and the subsequent creation of a central marketing authorization procedure with the emphasis of the approval process being on demonstration of quality, safety and efficacy of treatment. This involved: i) The creation of a committee for advanced therapies within the European Medicines Agency (EMA) to develop criteria and guidelines for product evaluation. ii) The provision of technical and risk management requirements to ensure quality, safety, efficacy, traceability and post-marketing surveillance. And iii) Special incentives to support innovation in Small-Medium sized Enterprises (SMEs) seeking to bring cell based therapeutics to market such as fast-track assessment and orphan drug status. Alternatives to the centralized route for advanced cell therapies include "hospital exemptions" where the therapy is experimental and patient specific and "specials" whereby the product is unlicensed and the protocol is modified to suit individual patients. These may prove to be the route by which new stem cell therapies founded on pluripotent stem cells are progressed in the short term in Europe. Against this centralized structure, executive oversight of the EUTCD directives is achieved via national authorities which ensure compliance with directives that have been incorporated into national statutes. These can vary between nation states. In the UK, regulatory oversight for compliance with the procurement, processing, storage and export of human tissues for clinical application is the responsibility of the Human Fertilisation Embryology Authority (HFEA) for gametes and embryos and the Human Tissue Authority (HTA) for all other tissues. Oversight of the inspection process for manufacture of advanced cellular therapies is the responsibility of the Medicines Healthcare Products Regulatory Agency (MHRA) (Fig. 7.1) with the EMA authorizing the necessary product licenses.

Reducing the risks of infectious disease transmission has been central to all EC directives and the guidelines on which they have been based. This is reflected by the emphasis placed on health history and status of prospective tissue donors and traceability of all tissues and underpinning reagents.

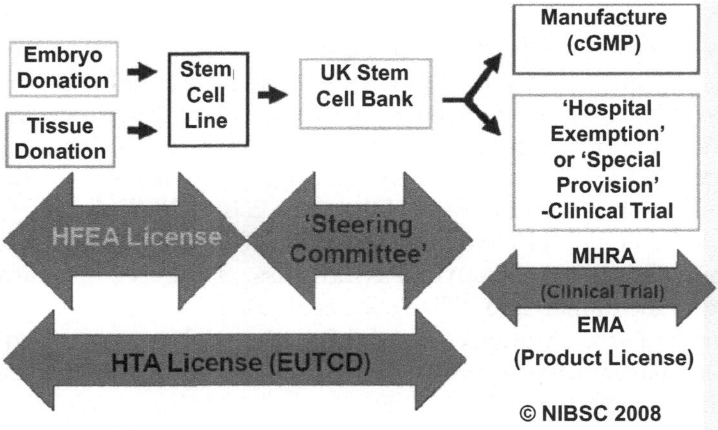

Figure 7.1. Overview of the statutory and non-satutory bodies involved in regulation of stem cell lines intended for human application. The body labelled "Steering Committee" is the Steering Committee for the UK Stem Cell Bank and for the Use of Stem Cell Lines which provides non-statutory oversight on the use of hESC lines. In particular its aim is to provide assurance to the public that the procurement of embryos, cell line derivation and application of any such lines used in the UK has been carried out in compliance with the UK requirements on ethics or under an equivalent system where the cells were derived outside of the UK. The jurisdictions of the HTA and HFEA overlap in the procurement of human reproductive tissues, embryos and gametes and consent but the two authorites are coordinated to avoid double inspection of certain procedures. One further element in UK regulation of stem cell therapies is the Gene Therapy Advisory Committee (GTAC, http://www.dh.gov.uk/ab/GTAC) which is the national ethics review body for clinical trials involving stem cells. For an overview of these regulations see the interim UK road map for stem cell therapy at http://www.advisorybodies.doh.gov.uk/genetics/gtac/interimukscroutemap120309.pdf and for step by step guidance see the respective E-Toolkit at http://www.sc-toolkit.ac.uk/regulatoryroutes/projectview.

While it is easy to envisage that the therapeutic use of cells in general or with hESC derivatives specifically could be achieved more expediently without these standards, the consequences associated with inconclusive or adverse results that could occur without having them could be dire. This is borne out by experience with gene therapy trials, or transmission of infectious pathogens via blood products or tissues (Gode and Bhide 1988, Simonds et al. 1992, MacLaren et al. 2005, Ludlam et al. 2006). While some of the complications that have been experienced could arguably be described as unforeseen on the basis of available knowledge, this in itself does not substantiate less rigorous standards. Rather, it emphasizes the need for the forethought that such standards invoke.

By comparison in the US, the US FDA regulates the use of human cells in human applications under *Title 21 Code of Federal Regulations (CFR)*, notably *parts 1270 (Human Tissue Intended for Transplantation)* and *1271 (Human Cell, Tissue, and Cellular and Tissue based Products*-**HCT/Ps**). These documents and other information on tissue related subjects are available at http://www.

fda.gov/cber/tissue/docs.htm, and http://www.fda.gov/cber/tiss.htm. In practice, the FDA's approach to cell and tissue regulation is risk based, and focused on three general areas: i) Limiting the risk of transmission of communicable disease from donors to recipients; ii) Establishing manufacturing practices that minimize the risk of contamination; and iii) Requiring appropriate demonstration of safety and effectiveness for cells and tissues that present greater risks due to their processing or their use. All HCT/Ps are regulated solely under Section 361 of the Public Health Services Act except for those requiring processing, which would include hESCs. These also require a premarket review and marketing license as an Investigational New Drug (**IND**). Current FDA and EMA guidelines for preparation of HCT/Ps are similar (see: http://www.fda.gov/Cber/gdlns/gtindcmc.htm; and http://www.emea.europa.eu/pdfs/human/cpwp/41086906en.pdf). However, there are differences in outlook on tissue sources (i.e., animal or human) and qualification of production reagents. To minimize opportunities for cross-specific disease transmission (i.e., zoonosis), the EMA recommends avoidance of animal sourced reagents and replacement with non-animal derived reagents of defined composition when they exist. By contrast, the FDA emphasizes the use of reagents from FDA-approved or cleared sources, or of clinical (pharmaceutical grade), irrespective of the species from which they are derived. To increase the likelihood of international utility it is therefore imperative to utilize Xeno-free reagents sourced from FDA qualified suppliers that also satisfy EMA requirements.

As exemplified by the difference between US and EU oversight of human cellular therapies it is vital for those seeking to develop human pluripotent stem cell based therapies, including stem cell banks, to be aware of which regulatory framework there are seeking to comply with.

Principles of good practice for evaluation of human pluripotent stem cell derived products in clinical trials

As the development of pluripotent stem cell based therapies will still take some time to mature, future-proofing of current standards is essential to ensure that today's investment will comply with evolving regulatory standards and meet with therapeutic success. Many of the same factors will need to be defined and optimized as currently met by transplantation of foetal and adult cells and tissues. This includes definition of the optimum cell lineage for transplantation, acceptable limits on non therapeutic cells which may be harboured in the graft, the route and timing of administration, "dosage" of cells and whether single or multiple administrations are required, definition of clinical subsets of disease, interplay of other medications and so on. Appropriate investigatory and clinical end points will need to be defined, as well as the necessity and ethical appropriateness of randomizing or blinding the study (Cesaro 2004). Basic cell culture principles are also important to address all such developments such as those published by (Coecke et al. 2005).

The extensive dependence of human pluripotent stem cell culture systems to expand and differentiate them, currently achieved by reliance on research grade or animal sourced reagents, arguably makes this a greater issue for the therapeutic use of these cells than for cells sourced from other tissues. Reagents need to be risk assessed, traceable and quality assured. Exposure to animal cell products that could theoretically brand a stem cell product as a xenobiotic should be minimized. This potentially could impact on patient follow-up in a clinical trials. An additional key feature that must be high priority for stem cell banks is the documentation needed to show traceability of processes, environmental conditions and sources of cells and reagents. It is unlikely that culture processes will be perfect as the first trials begin. Reagents of animal origin may persist as they may well remain key to cellular performance and as always a balance will be struck between risk and benefit. Inevitably this may mean that early clinical applications will focus on therapies for patients who are gravely ill with poor prognosis. However, as experience develops improvements in the culture process will come to enable trials in a broader range of patients.

The capacity of pluripotent stem cells to form tumours and patient immune response against transplanted stem cell derived products are two other important issues that need to be addressed in evaluating the safety of pluripotent stem cells for therapy. Clearly the ability to assure the absence of undifferentiated, and potentially tumourigenic, cells in the final product will be vital to regulatory acceptability. Products will also have to exhibit robust and long term avoidance of the recipient patient's immune responses. Current approaches include the application of stem cells in immune privileged sites such as the eye and central nervous system, and the use of cytotoxic drugs. However, it is yet to be confirmed whether such approaches will yield long term therapeutic success and other developments, possibly including the availability of banks of pluripotent cell lines of diverse genotypes to enable a suitable match with patients, may be required.

Current clinical applications using mesenchymal stem cell cultures could also provide valuable experience using certain reagents such as growth factors and culture matrices. In the EU/UK, clinical trials of experimental therapeutics need to be carried out in compliance with 1) the EU Directive on clinical trials (2001/20/EC), and 2) the Research Governance Framework for Health and Community Care. These cover the responsibilities of the Sponsor, approval of the trial by the relevant competent authority (in the UK, the MHRA) and Ethics Committee(s) and the conduct of the study according to the principles of Good Clinical Practice. In addition, any clinical trial must be reviewed and monitored by an independent trial monitoring committee. Any experimental therapeutic product would be manufactured in compliance with Best Practice standards and systems would need to be in place for the recording and investigation of any adverse events occurring during the study and ongoing pharmacovigilance. Written informed consent would be obtained for all participants before commencement on the study and specific documented procedures would be in place for the handling of donor and patient trial samples and archive material. Patient and donor related data would be handled in accordance with the Data Protection Act 1998 and strict confidentiality would be maintained. The issue of compensation in the event of non-negligent harm would be addressed in line with Ethics Committee requirements and appropriate indemnity established.

Conclusion

Over 30 yr from the first isolation of pluripotent stem cells from mouse embryos, and more than a decade from a similar achievement in humans, the world is set to reap the translational benefits of this achievement. These will range from the industrial screening applications to augment existing strategies for drug discovery or toxicology, and as a source of cells for therapeutic intervention in regenerative medicine. To facilitate this substantial progress has been made in the development of culture systems to grow these cells and preserve their properties, founded on research to understand the molecular mechanisms which define their phenotype. This knowledge and further advances in defining the composition of *in vitro* systems for both growth and storage and high throughput systems to automate their production will be critical to achieve their efficacious production, qualification and safety for human application. The latter will require compliance of product specification with regulatory standards which in Europe are undergoing a welcome process of harmonization. As such the future is bright for human pluripotent stem cells as products for regenerative medicine to provide new solutions for the treatment of disease and tissue damage. While the road has been long, good things normally do come to those who wait.

Acknowledgements

The authors thank Ms. Leigh Murray for assistance with formatting of the manuscript.

References

Allegrucci C, YZ Wu, A Thurston et al. 2007. Restriction landmark genome scanning indentifies culture-induced DNA methylation instability in the human embryonic stem cell epigenome. Hum Mol Genet. 16: 1253–1268.

Andrews PD, M Becroft, A Aspegren et al. 2010. High content screening of feeder free human embryonic stem cells to identify pro-survival molecules. Biochem J. 432(1): 21–33.

Amit M, MK Carpenter, MS Inokuma et al. 2000. Clonally derived human embryonic stem cell lines maintain pluripotency and proliferative potential for prolonged periods of culture. Dev Biol. 227: 271–8.

Avery S, G Zafarana, PJ Gokhale et al. 2010. The role of SMAD4 in human embryonic stem cell self-renewal and stem cell fate. Stem Cells. 28(5): 863–73.

Ben-David U, N Benvenisty and Y Mayshar. 2010. Genetic instability in human induced pluripotent stem cells: classification of causes and possible safeguards. Cell Cycle. 1; 9(23): 4603–4.

Bendall SC, MH Stewart, P Menendez et al. 2007. IGF and FGF cooperatively establish the regulatory stem cell niche of pluripotent human cells *in vitro*. Nature. 2007 Aug 30; 448(7157): 1015–21.

Bibikova M, E Chudin, B Wu et al. 2006. Human embryonic stem cells have a unique epigenetic signature. Genome Res. 16: 1075–1083.

Burton P, DR Adams, A Abraham et al. 2010. Erythro-0-(2-hydroxy-3-nonyl)adenine (EHNA) blocks differentiation and maintains the expression of pluripotency markers in human embryonic stem cells. Biochemical Journal. 432(3): 575–584.

Catalina P, C Bueno, R Montes et al. 2009. Genetic stability of human embryonic stem cells: A first step toward the development of potential hESC-based systems for modelling childhood leukaemia. Leuk Res. 33(7): 980–990.

Cesaro P. 2004. The design of clinical trials for cell transplantation into the central nervous system. NeuroRx. 1: 492.

Chen S, J-T Do, Q Zhang et al. 2006. Self-renewal of embryonic stem cells by a small molecule. PNAS. 103: 17266–71.

Cho YM, S Kwon, YK Pak et al. 2006. Dynamic changes in mitochoudrial biogenesis and antioxidant enzymes during the spontaneous differentiation of human embryonic stem cells. Biochem Biophys Res Commun. 348(4): 1472–1478.

Coecke S, M Balls, G Bowe et al. 2005. Guidance on good cell culture practice. a report of the second ECVAM task force on good cell culture practice. Altern Lab Anim. 33(3): 261–87.

Costa FF, EA Seftor, JM Bischof et al. 2009. Epigenetically reprogramming metastatic tumor cells with an embryonic microenvironment. Epigenomics. 1(2): 387–398.

Damoiseaux R, SP Sherman, JA Alva et al. 2009. Integrated chemical genomics reveals modifiers of survival in human embryonic stem cells. Stem Cells. 27: 533.

Davis J. 2007. Systems for cell culture scale up. In: Medicines from Animal Cells. Stacey G and Davis J. J Wiley and Sons, Chichester, UK, pp. 145–172.

Desbordes SC, DG Placantonakis, A Ciro et al. 2008. High-throughput screening assay for the identification of compounds regulating self-renewal and differentiation in human embryonic stem cells. Cell Stem Cell. 2: 602.

De Sousa PA, G Galea and M Turner. 2006 The road to providing human embryo stem cells for therapeutic use: the UK experience. Reproduction. 132: 681–9.

Fletcher JM, PM Ferrier, JO Gardner et al. 2006. Isolation of new human embryonic stem cell lines without direct exposure to animal cell products. Cloning and Stem Cells. 8 (No. 4): 319–334.

Forsyth NR, A Kay, K Hampson et al. 2008. Transcriptome alterations due to physiological normoxic (2% O2) culture of human embryonic stem cells. Regen Med. 3: 817–833.

Fujioka T, K Yasuchika, Y Nakamura et al. 2004. A simple and efficient cryopreservation method for primate embryonic stem cells. The International Journal of Developmental Biology. 48(10): 1149–1154.

Gerecht S, JA Burdick, LS Ferreira et al. 2007. Hyaluronic acid hydrogel for controlled self-renewal and differentiation of human embryonic stem cells. Proc Natl Acad Sci USA. 104 (27): 11298–11303.

Gode GR and NK Bhide. 1988. Two rabies deaths after corneal grafts from one donor. Lancet 2 791.

Hendrix MJ, EA Seftor, RE Seftor et al. 2007. Reprogramming metastatic tumour cells with embryonic microenvironments. Nat Rev Cancer. 7(4): 246–255.

Heng BC, SM Bested, SH Chan et al. 2005. A proposed design for the cryopreservation of intact and adherent human embryonic stem cell colonies. *In Vitro* Cellular & Developmental Biology. 41(3-4): 77–79.

Holubcová Z, P Matula, M Sedláčková et al. 2011. Human embryonic stem cells suffer from centrosomal amplification. Stem Cells. 29(1): 46–56.

Hunt CJ. 2011. Cryopreservation of human stem cells for cinical application: a review. Transfusion Medicine and Hemotherapy. 38(2): 107–123.

Inglis SR, JJ Turner and MM Harding. 2006. Applications of type I antifreeze proteins: studies with model membranes & cryoprotectant properties. Curr Protein Pept Sci. 7(6): 509–22.

International Stem Cell Initiative. 2007. Characterisation of human embryonic stem cell lines by the International Stem Cell Initiative. Nature Biotechnology. 25: 803–816.

International Stem Cell Initiative. 2010. Comparison of defined culture systems for feeder cell free propogation of human embryonic stem cells. *In vitro* Cell Dev Biol Anim. 46: 247–258.

King FW, C Ritner, W Liszewski et al. 2009. Subpopulations of human embryonic stem cells with distinct tissue specific fates can be selected from pluripotent cultures. Stem Cells Dev. 18(10): 1441–1450.

Krawetz R, JT Taiani, S Liu et al. 2009. Large-scale expansion of pluripotent human embryonic stem cells in stirred suspension bioreactors. Tissue Eng Part C Methods. 16(4): 573–82.

Laurent LC, I Ulitsky, I Slavin et al. 2011. Dynamic changes in copy number of pluripotency and cell proliferation genes in human ESCs and iPSCs during reprogramming and time in culture. Cell Stem Cell. 8(1): 106–118.

Lane M, JM Maybach, K Hooper et al. 2003. Cryo-survival and development of bovine blastocysts are enhanced by culture with recombinant albumin and hyaluronan. Molecular Reproduction and Development. 64(1): 70–78.

Lefort N, M Feyeux, C Bas et al. 2008. Human embryonic stem cells reveal recurrent genomic instability at 20q11.21. Nat Biotechnol. 26(12): 1364–1366.

Li Y, S Powell, E Brunette et al. 2005. Expansion of human embryonic stem cells in defined serum-free medium devoid of animal-derived products. Biotechnol Bioengin. 91: 688–698

Li YJ, EH Chung, RT Rodriguez et al. 2006. Hydrogels as artificial matrices for human embryonic stem cell self-renewal. J Biomed Mater Res. 79(1): 1–5.

Lu J, R Hou, CJ Booth et al. 2006. Defined culture conditions of human embryonic stem cells. Proc Natl Acad Sci USA. 103: 5688.

Ludlam CA, WG Powderly, S Bozzette et al. 2006. Clinical perspectives of emerging pathogens in bleeding disorders. Lancet. 367: 252–61.

Ludwig TE, ME Levenstein, JM Jones et al. 2006. Derivation of human embryonic stem cells in defined conditions. Nat Biotechnol. 24: 185.

MacLaren RE, RR Ali and AJ Thrasher. 2005. Risks of gene therapy should be weighed against lack of alternatives for many diseases. BMJ. 330: 791.

Maitra A, DE Arking, N Shivapurkar,et al. 2005. Genomic alterations in cultured human embryonic stem cells. Nat Genet. 37(10): 1099–103.

Makoschey B, PT van Gelder, V Keijsers et al. 2003. Bovine viral diarrhoea virus antigen in foetal calf serum batches and consequences of such contamination for vaccine production. Biologicals. 31(3): 203–8.

Miyabayashi T, JL Teo, M Yamamoto et al. 2007. Wnt/β-catenin/CBP signaling maintains long-term murine embryonic stem cell pluripotency. Proc Natl Acad Sci USA. 104(13): 5668–5673.

Mumaw JL, D Machacek, JP Shields et al. 2010. Neural differentiation of human embryonic stem cells at the ultrastructural level. Microsc Microanal. 16(1): 80–90.

Närvä E, R Autio, N Rahkonen et al. 2010. High resolution DNA analysis of human embryonic stem cell lines reveals culture induced copy number changes and loss of heterozygosity. Nat Biotechnol. 28(4): 371–377.

Oh SK, HS Kim, HJ Ahn et al. 2005. Derivation and characterisation of new human embryonic stem cell lines: SNUhES1, SNUhES2, and SNUhES3. Stem Cells. 23(2): 211–219.

Oh WKW, AK Chen, Y Mok et al. 2009. Long term microcarrier suspension cultures of human embryonic stem cells. Stem Cell Res. 2(3): 219–30.

Park SH, SH Park, MC Kook et al. 2004. Ultrastructure of human embryonic stem cells and spontaneous and retinoic acid-induced differentiating cells. Ultrastruct Pathol. 28(4): 229–238.

Phillips BW, R Horne, TS Lay et al. 2008. Attachment and growth of human embryonic stem cells on microcarriers. J Biotechnology. 138: 24–32.

Placzek MR, I-M Chung, H Macedo et al. 2009. Stem cell bioprocessing : fundamentals and principles. J Royal Society Interface. 6: 209–232.

Postovit LM, NV Margaryan, EA Seftor et al. 2008. Human embryonic stem cell microenvironment suppresses the tumorigenic phenotype of aggressive cancer cells. Proc Natl Acad Sci USA. 105(11): 4329–4334.

Rao BM and PW Zandstra. 2005. Culture development for uman embryonic stem cell propagation: molecular aspects and challenges. Current Opinion in Biotechnology. 16: 568–576.

Richards M, SP Tan, JH Tan et al. 2004. The transcriptome profile of human embryonic stem cells as defined by SAGE. Stem Cells. 22(1): 51–64.

Roobrouck VD, K Vanuytsel and CM Verfaillie. 2011. Concise review: culture mediated changes in fate and/or potency of stem cells. Stem Cells. 29(4): 583–589.

Reubinoff BE, MF Pera, G Vajta et al. 2001. Effective cryopreservation of human embryonic stem cells by the open pulled straw vitrification method. Human Reproduction (Oxford, England). 16(10): 2187–2194.

Sathananthan AH and A Trounson. 2005. Human embryonic stem cells and their spontaneous differentiation. Ital J Anat Embryol. 110(2 Suppl 1): 151–157.

Sato N, L Meijer, L Skaltsounis et al. 2004. Maintenance of pluripotency in human and mouse embryonic stem cells through activation of Wnt signalling by a pharmacological GSK-3-specific inhibitor. Nature Medicine. 10: 55–63.

Simonds RJ, SD Holmberg, RL Hurwitz et al. 1992. Transmission of human immunodeficiency virus type 1 from a seronegative organ and tissue donor. N Eng J Med. 326: 726–732.

Siti-Ismail N, AE Bishop, JM Polak and A Mantalaris. 2008. The benefit of human embryonic stem cell encapsulation for prolonged feeder-free maintenance. Biomaterials 2008 Oct; 29(29): 3946–52: Epub 2008 Jul 17.

Sovalat H, M Scrofani, A Eidenschenk et al. 2011. Identification and isolation from either adult human bone marrow or G-CSF–mobilized peripheral blood of $CD34^+/CD133^+/CXCR4^+/Lin^-CD45^-$ cells, featuring morphological, molecular, and phenotypic characteristics of very small embryonic-like (VSEL) stem cells. Exp Hematol. 39(4): 495–505.

Spits C, I Mateizel, M Geens et al. 2008. Recurrent chromosomal abnormalities in human embryonic stem cells. Nat Biotechnol. 26(12): 1361–1363.

Stewart MH, SC Bendall, M Levadoux-Martin et al. 2010. Clonal tracking of hESCs reveals differential contribution to functional assays. Nat Methods. 7(11): 917–922.

Stojkovic M, S Kölle, S Peinl et al. 2002. Effects of high concentrations of hyaluronan in culture medium on development and survival rates of fresh and frozen-thawed bovine embryos produced *in vitro*. Reproduction. 124(1): 141–153.

Suemori H, K Yasuchika, K Hasegawa et al. 2006. Efficient establishment of human embryonic stem cell lines and long-term maintenance with stable karyotype by enzymatic bulk passage. Biochemical and Biophysical Research Communications. 345(3): 926–932.

Takahashi K and S Yamanaka. 2006. Induction of pluripotent stem cells from mouse embryonic and adult fibroblast cultures by defined factors. Cell. 126: 663–676.

Tanasijevic B, B Dai, T Ezashi et al. 2009. Progressive accumulation of epigenetic heterogeneity during human ES cell culture. Epigenetics. 4(5): 330–338.

Thomas RJ, D Anderson, A Chandra et al. 2009. Automated, Scalable Culture of Human Embryonic Stem Cells in Feeder Free Conditions. Biotechnol Bioeng. 102(6): 1636–1644.

Thomson JA, J Itskovitz-Eldor, SS Shapiro et al. 1998. Embryonic tem cell lines derived from human blastocysts. Science. 282: 1145–1147.

Tsutsui H, B Valamehr, A Hindoyan et al. 2011. An optimised small molecule inhibitor cocktail supports long-term maintenance of human embryonic stem cells. Nat Commun. 2: 167.

Vanderzwalmen P, G Bertin, CH Debauche et al. 2003. Vitrification of human blastocysts with the Hemi-Straw carrier: application of assisted hatching after thawing. 18(7): 1504–1511.

Wang L, TC Schulz, ES Sherrer et al. 2007. Self-renewal of human embryonic stem cells requires insulin-like growth factor-1 receptor and ERBB2 receptor signalling. Blood. 110(12): 4111–4119.

Werbowetski-Ogilvie TE, M Bossé, M Stewart et al. 2009. Characterisation of human embryonic stem cells with features of neoplastic progression. Nat Biotechnol. 27(1): 91–97.

Xiao L, X Yuan and SJ Sharkis. 2006. Activin A maintains self-renewal and regulates fibroblast growth factor, Wnt, and bone morphogenic protein pathways in human embryonic stem cells. Stem Cells. 24(6): 1476–1486.

Xu C, MS Inokuma and J Denham et al. 2001. Feeder-free growth of undifferentiated human embryonic stem cells. Nat Biotechnol. 19(10): 971–4.

Xu RH, RM Peck, DS Li et al. 2005. Basic FGF and suppression of BMP signalling sustain undifferentiatied proliferation of human ES cells. Nature Methods. 2: 185–90.

Xu Y, X Zhu, HS Hahm et al. 2010. Revealing a core signalling regulatory mechanism for pluripotent stem cell survival and self renewal by small molecules. Proc Natl Acad Sci USA. 107(18): 8129–8134.

Yao S, S Chen, J Clark et al. 2006. Long-term self-renewal and directed differentiation of human embryonic stem cells in chemically defined conditions. Proc Natl Acad Sci US. 103, 6907.

8

Cells for Bioartificial Liver Support

Humphrey Hodgson,[1,a,*] ***Amy Thomas***[1,b]
and *Clare Selden*[1,c]

The quest for an artificial liver has a long and thus far unsuccessful history. Tygstrup (Christensen et al. 1984) defined the zone of clinical practice in which liver replacement might be helpful, buying time for the liver to recover by repair and regeneration. It would be unnecessary in those who will recover spontaneously; it would be futile if there is no prospect of recovery of native liver function due to the severity of liver disease, co-morbidities or irreversible liver failure; it could however have a dramatic effect in patients lying between these two extremes. The availability of organ transplantation extended the potential scope, as a clinical picture with reversible liver function is not required. However in irreversible failure, as there is often a delay before organs can be found, "bridging to transplant" became an important concept for the utility of liver support. An important analogy from current clinical practice was provided by the rare but effective procedure of auxiliary liver transplantation followed by withdrawal of immune suppression: this procedure demonstrated in the clearest terms that transient biological liver support added at the time of severe life-threatening liver failure can buy time for repair and regeneration of the native liver.

[1]Royal Free Campus, UCL Medical School, London NW3 2PF.
[a]Email: hhodgson@ucl.ac.uk
[b]Email: athomas@medsch.ucl.ac.uk
[c]Email: cselden@medsch.ucl.ac.uk
*Corresponding author

A considerable number of approaches to design a liver replacement have been tried, ranging from artificial devices based on physical principles such as dialysis or adsorption columns—to cell based devices, and extracorporeal perfusion through whole livers of animal or human origin. There have recently (2010) been reported multicentre trials of "state-of-the-art" non-biological liver machines, which have not improved mortality (Rifai et al. 2010, Banares et al. 2010). We and others have concluded that the functions of the liver are so complex and numerous (detoxification, and synthesis of many different substances) that only a bioartificial liver (BAL), using living cells expressing liver-specific functions, can substitute liver function effectively.

Brief reference must be made to the use of whole organs as biological support. There are historical reports of cross-perfusion experiments in man, and a relatively small number of reports in the literature on the use of explanted whole human livers in extracorporeal circuits (Horslen et al. 2000). Whilst they appear to have been effective short term, the competition between the transplant surgeon who wishes to use an explanted liver in a procedure of proven effectiveness, and the experimental use of the organ in an extracorporeal device, in which its utility is very short lived and the organ is lost from the transplant pool, has not encouraged the development of this technique. The use of animal (baboon, pig) livers, though reported with occasional success, has floundered on the biological incompatibilites of xenogeneic tissues (Klebanoff et al. 1973). Genetically modified pigs as a source of non-immunogenic animal organs remains a distant future prospect—the possibilities are largely being explored in the area of cardiac transplantations, but the development of bioartificial livers (by for example the inhibition of complement activation) is an attractive option (Matsushita et al. 2003). However, even if the hyper-acute immune responses that might follow the exposure of pig cells to human blood or serum can be obviated, doubts will remain about functional protein-protein incompatibilities between porcine and human serum components.

Bioartificial livers

Whilst maintaining whole organs extra-corporeally without rapid and progressive loss of function is extremely difficult, maintaining cells in culture is a highly developed technology. Applying this to extracorporeal liver support the key questions are: Which liver cells? How many? From which source? Cultured in what form? Maintained in what kind of extracorporeal chamber (bioreactor)? This chapter concerns the potential cell sources and types that are or could become available, and the problems presented by securing an adequate supply.

Which liver cells?

The liver's critical functions reside in the hepatocyte, 60% of the liver's cell population numerically, 90% by mass. Hepatocytes synthesize and secretes proteins, and detoxify endogenous and exogenous molecules. Therefore replacing hepatocyte function is the prime aim of a BAL. Most bioartificial livers do not look past this sub-population, although there are significant contributions in respect of cytokine production (notably from Kupffer cells and also endothelial cells) and growth factor production (notably from stellate cells) that must be missing from hepatocyte-only machines. There are also in the native liver key inter-relationships by which hepatocytes maintain polarity and function by their relationship to other cells. It has also been argued that liver cells in an extracorporeal circuit will not function adequately as they lack a biliary tree which normally eliminates many detoxified products. However, the biliary tree is a collecting duct for substances rendered water-soluble during passage through hepatocytes; provided liver cells in a BAL are polarized, with a canalicular membrane, water-soluble detoxified products will access the patient's circulation for final excretion through the kidney. Currently therefore the provision of an adequate mass of hepatocytes remains the prime objective of a bioartificial liver. Such a mass must also express the key detoxificatory and synthetic functions required, and this aspect is discussed below as each potential cell source is discussed.

How many?

In man in the most extreme instance survival has occurred after 85–90% hepatectomy (Starzl et al. 1975), but a more general figure for surgical resection is a minimal functional residual liver volume following extended hepatectomy of 25% or more in a normal liver, and 40% or more with preoperative liver dysfunction. How do these figures relate to the number of cells required in a bioartificial liver? The hepatocyte number in the adult liver is of the order of 2×10^{11}. The function of the cells in a bioartificial liver will be additive to the function still expressed by the patient's own damaged liver. Adding 10% functionality might make the critical difference —but unless each hepatocyte in the bioreactor functions as efficiently as *in vivo,* more than 10^{10} cells will be needed. Our own calculations and bioreactor specifications provide for a cell number of 7×10^{10}, 30% of the native hepatocyte population. Certainly the much smaller cell numbers, of the order of 5×10^{9}, used in the xenogeneic pig hepatocyte system that failed in clinical trial seem conceptually inadequate (Demetriou et al. 2004).

From which source?

Whilst primary human hepatocytes might appear the ideal cell culture, this is not so in practice. Firstly they are not readily available—explanted organs that could provide the cells are diverted to transplant surgeons (or in small numbers to cell transplantation). Furthermore, primary human hepatocytes in culture have two significant disadvantages: they cannot currently be induced to proliferate in culture to provide a reliable and predictable cell source for filling bioreactors, and primary hepatocytes tend rapidly to de-differentiate in culture unless maintained with complex matrices that may be unsuitable for use in a perfused bioreactor. Thus although there have been bioartificial livers designed based on human hepatocytes, clinical usage thus far has been limited to case reports documenting single cases.

Accepting the drawbacks of the use of primary human hepatocytes, a number of paths have been explored a) non-human hepatocytes, b) engineered hepatocytes to allow proliferation, c) tumour-derived cell-lines which by definition proliferate d) foetal hepatocytes which maintain proliferation in culture, and e) cells matured from progenitor cells and/or stem cells. Each alternative has potential but also disadvantages.

Non-human hepatocytes

Whilst non-human hepatocytes are readily available—notably from pigs—there are serious concerns about protein-protein incompatibilities. Additionally, the carriage of pig endogenous retroviruses within the porcine genome raises the possibility that—at worst—a new retroviral threat to the human species could be unleashed. Limited evidence from patients exposed to pig cells has not revealed any evidence of such transmission, but moratoria have been introduced in various countries on their use, and *in vitro* systems have demonstrated that retroviral transmission to co-cultured cells can occur (Yang et al. 2010). Pig-based machines have indeed been used in clinical practice—most notably in the Cedars–Sinai machine based on hepatocytes cultured on microspheres. However this machine used only a liver mass of 5×10^9 cells, and thus its failure in a clinical trial cannot firmly be attributed to the use of xenogeneic cells, and is more likely to reflect an inadequate biomass (Demetriou et al. 2004). Similarly there are case reports of patients "bridged to transplantation" with—or during—the use of pig-based machines, but their value has not been established. They continue however to be the cell source in a considerable number of pre-clinical trials.

Engineered hepatocytes to allow proliferation

Engineering primary human hepatocytes—by gene transfection with for example SV40 large-t antigen or telomerase, has produced proliferating cell lines, but the performance of those cells compared with primary cells has been disappointing, and they have not yet been of sufficient value to be trialled clinically (Smalley et al. 2001, Wege et al. 2003). This exemplifies the central dilemma of proliferating cells—that the necessity for self-replication imposes a functional profile very distinct from the normal resting hepatocyte, and the key functions that are required for liver support, notably detoxification and the synthesis of non-structural proteins, are precisely those that are down-regulated during the proliferative phase.

In silico or *in vitro* assessment of potential proliferating liver cells for use in a bioartificial liver must address those functional areas—and the detoxification profile—notably Cytochrome P450 (CYP) activity and urea detoxification are clearly critical in this regard. CYP enzymes are for example, responsible for in excess of 90% of xenobiotic metabolism (Yan and Caldwell 2001). CYP3A4 is the most abundant CYP enzyme in adult human liver, governing around 45% of xenobiotic metabolism with the least substrate specificity—it is thus the most significant CYPP enzyme for a BAL. Recently, a number of immortalized adult human hepatocyte cell lines have been developed with CYP3A activity. These are not necessarily suited however to large scale BAL culture. HepaRG cells only differentiate into hepatocyte-like morphology following lengthy treatment with 2% DMSO (Kanebratt and Andersson 2008), and HC-04 and Fa2-N4 cell lines are both cultured on collagen matrices, each of which may lead to difficulties in obtaining regulatory approval for devices containing them. Furthermore P450 function of these cell lines may be insufficient. CYP3A4 protein expression can be increased in HC-04 cells by the PXR agonist, however their drug metabolizing capacity is equivalent only to adult primary hepatocytes from donors with low levels of CYP3A4 function (Fontana et al. 1999). Fa2-N4 cells demonstrate both CYP1A2 (EROD) and CYP3A4 (testosterone, midazolam) activity, but their levels of function are >75% lower than primary cells. Additionally, whilst levels of mediators of CYP1A2 (arylhydorcarbon receptor) and CYP2C/3A (PXR) induction in Fa2-N4 cells match those found in adult primary hepatocytes, their reduced drug metabolism capacity could partially be explained by a deficiency in the CYP2B/CYP3A mediator constitutive androstane receptor and by decreased expression of hepatic uptake transporters (Lim and Tan 2007, Hariparsad et al. 2008).

Tumour-derived cell-lines

To date the most available source of proliferating cells are those cell-lines initially derived from well-differentiated hepatocellular tumours, such as HepG2 cells. They are readily available and proliferate well in culture, theoretically therefore being available in unlimited quantities. However, as with all tumour-derived cells, proliferation is associated with a relative loss of some differentiated function. In general—and particularly if the culture conditions are optimized as we describe below—synthesis of many proteins can be achieved at *in vivo* equivalent levels, but there are important relative deficits particularly in detoxificatory functions. Transcripts of CYP3A4 are present in HepG2 cells although mRNA levels are far lower (100–1000 fold) than in adult primary human hepatocytes (Kenny et al. 2008) although there is a large inter-individual variability between cells from livers from different individuals (Wilkening et al. 2003, Donato et al. 2007). Interestingly our data demonstrated that microsomes extracted from Hep G2 cells, and also that HepG2 cells cultured in optimal matrix configurations, could show CYP 450 activity within the lower tertile of that of a panel of human primary human hepatocytes.

Because decreased mRNA expression of CYPs appears responsible for the low functional activities in hepatoma cells (Rodriguez-Antona et al. 2002), rectifying expression of CYP mRNA should improve enzyme function. HepG2 cell lines stably expressing single isoforms of CYPs have been engineered (Araki et al. 2008, Yoshitomi et al. 2001) but sustained function of multiple CYP isoforms at a level high enough for use in a BAL is difficult to achieve. A combination of chemical induction and genetic manipulation might be a more successful approach. Alternative cell lines could also be considered. The intestine is an expression site of CYP3A4/5, intestinal cell lines (Harmsen et al. 2008) could therefore be cocultured alongside a liver cell line to provide the detoxificatory component of a BAL, but as CYP3A function in intestinal cell lines is low in comparison to primary human hepatocytes they probably have no advantage over cell lines of hepatic origin.

Another major area of detoxificatory concern, in respect of common tumour-derived cell lines, is their ability to metabolize ammonia. It is well established that hyperammonaemia occurs in liver failure (e.g., plasma ammonia levels elevated from 10–47umol/L to 134–188umol/L in acute liver failure (Kundra et al. 2005)) and this appears a major driver towards the development of hepatic encephalopathy. Ammonia elimination is thus an essential feature for achieving a functional and applicable liver support system. Efforts have been made to eliminate ammonia from patient's plasma by use of ion exchange, adsorption and diffusion in artificial liver support systems, but although reductions in circulating ammonia levels can be achieved in cell-free systems, their clinical applicability is less clear.

In vivo ammonia disposal is via urea cycle which comprises five enzymes present in two different cellular compartments, namely carbamyl phosphate synthetase-I (CPS-I) and ornithine transcarbamylase (OTC) present in mitochondria, and argininosuccinate synthetase (ASS), argininosuccinate lyase (ASL), and arginase (Arg) are present in the cytoplasm of the cell respectively. The enzyme Arginase exist in two forms and in different cellular compartments, Arginase I predominantly present in the mitochondria and Arginase II largely present in the cytoplasm. Most studies suggest that Arginase II is more commonly expressed in cancerous cells (Cederbaum et al. 2004). Urea production reported in cell culture may often represent an Arginase II pathway rather than Arginase I, and the former pathway does not contribute to ammonia detoxification.

Most liver cell lines which are candidates to form the biomass of BAL lack a fully functional urea cycle, and when assessed for ammonia elimination perform poorly compared to liver. An in depth analysis of the mRNA analysis of three mammalian cell lines cBAL111, C3A and HepG2 revealed that these potential cell line lack the expression of two genes OTC and Arginase I (Mavri-Damelin et al. 2008, Mavri-Damelin et al. 2007), even when cells performance was optimized using a 3D culture.Coordinated multi gene expression of the deficient enzymes could re-instate the urea cycle. Mavri-Damelin et al. (2007) expressed both OTC and ArgI in HepG2 cell line concomitantly, leading to a rise in urea production, thus establishing proof of principle; however the urea production was significantly lower than primary hepatocytes and liver, and moreover the expression fell away on cell passage. Achieving multigene expression has been challenging to vector biologists, but efficient multicistronic vectors capable of coordinated gene expression of OTC and ArgI under constitutive promoters functional in liver cells would be the key to fully functional urea cycle restoration and provision of a valuable source of biomass essential for BAL.

Foetal hepatocytes

Foetal hepatocytes (FHH) do proliferate, and can be further manipulated to "immortalization", and express relevant CYPs, and could thus provide P450 function. However P450 function of an immortalized foetal hepatocyte cell line cBAL111 was only 0.1% that levels measured in adult PHH, and disappointingly whilst ammonia elimination was achieved this was towards glutamate synthesis rather than via the urea cycle which may in fact precipitate or worsen the hepatic encephalopathy (Poyck et al. 2008). However FHH also differ from adult hepatocytes not only in their preferential expression of CYP3A7 to CYP3A4 but also in their negligible expression of pregnane X receptor (PXR) (Maruyama et al. 2007, Matsunaga et al. 2004) which is a key mediator of CYP3A induction. Indeed P450 function in passaged foetal hepatocytes is lower than in isolated adult primary human hepatocytes and although higher CYP activity can be achieved by clonal selection, extensive *in vitro* expansion of FHH may result in decreased hepatic function (Deurholt et al. 2006). Thus obtaining an adequate biomass comprised of foetal hepatocytes could be problematic.

Comparisons

There have been relatively few "head-to-head" comparisons of different cell types, even *in vitro*. The Amsterdam group compared the efficacy of human foetal, human adult, and porcine adult primary hepatocytes (Poyck et al. 2007). The human adult cells eliminated 2-fold more ammonia and produced 3-fold more urea than the porcine, whilst the foetal hepatocytes in fact produced ammonia. Lidocaine elimination of by the foetal cells, was however, significantly higher than either adult cells type. Albumin production was not different between the three cell types, and metabolically the adult human cells were the most stable. Thus, disappointingly for practical considerations, they judged the normal mature primary human hepatocyte as the optimal cell type.

Cells matured from progenitor cells and/or stem cells

The greatest area of excitement concerning future technologies lies in the use of progenitor cells. As already discussed, the adult liver is relatively mitotically inactive, though not absolutely quiescent, and a liver cell progenitor was difficult to identify, particularly as in severe disease it is the mature hepatocytes that appear to provide the main proliferative compartment. This ability to proliferate readily and rapidly *in vivo* contrasts the inability to act similarly *in vitro*. However it is now clear that there is within the liver a stem cell—indeed bipotential with the ability to proliferate and mature into either hepatocytes or biliary epithelial cells—located at the critical junction between hepatocytes and biliary canaliculi, the Canal of Hering (Kuwahara et al. 2008). Indeed there is recent compelling evidence that this may well be responsible for a very slow turnover of hepatocytes in normal liver. It is also now clear that in disease, when the proliferative capacity of the adult hepatocyte population is impaired, this compartment can expand rapidly producing small "oval cells" which can mature into hepatocytes.

Considerable literature also emerged over the last decade concerning non-hepatic cells that could migrate to the liver and mature into hepatocytes. Whilst it is now apparent that in the majority of instances cited this was not transdifferentiation but cell fusion between the putative stem-cell and resident hepatocytes within the liver, there are some well-documented examples where this phenomenon is indeed transdifferentiation (Dorrell and Grompe 2005). The cell sources include bone marrow, peripheral stem cells mobilized from the marrow, umbilical cord blood and the amnion, and indeed the macrophage population of peripheral blood. Investigating the hepatic potential of non-hepatic cells has lead to characterization of a number of cell surface markers indicating this potentiality, including notably CD133, Thy1, EPCAM—which differ to some extent between mouse and human systems—as well as developments in both mRNA and protein and functional assessment of their maturation state. Key messages that have emerged in this regard are the use of AFP as a marker of early differentiation towards hepatocyte function; description of later markers of more mature hepatocyte function such as tyrosine aminotransferase; and perhaps most importantly recognition that mRNA expression for hepatocyte-specific proteins is not necessarily accompanied by significant protein expression and function—and certainly not at the levels that would be required to replicate hepatocyte function in a bioartificial liver.

Compared to cells derived from other organs, embryonic stem cells, in the form of available hESC lines, offered until recently the most obvious source to exploit if they can effectively be matured to hepatocytes (excluding the ethical debate which the author does not propose to enter). They have the advantage that there are well characterized cell lines available, providing effectively unlimited quantities of cells. Thus perfecting maturation protocols presents fewer problems than those which involve—effectively many times over—using non-uniform starting material of cells derived from sources such as bone-marrow or amnion.

One of the characteristics of an embryonic stem cell is its potential for tri-lineage differentiation, most strikingly demonstrated by *in vivo* teratoma formation after implantation into nude mice. However whilst the teratogenic potential of hESCs has a major impact on their potential use *in vivo*, it scarcely pertains in the setting of usage in an extracorporeal circulation, where a system of cell filters can prevent cell fragments and cell-derived DNA entering the circulation of a patient.

Recently, the ground breaking studies which demonstrated the derivation of induced pluripotent stem cells (IPSCs) from fibroblasts, by genetic manipulation of initially four (Oct3/4, Sox2, c-Myc, and Klf4, or Oct3/4, Sox2, Nanog, and Lin28) genes has provided the field with an excellent and tissue matched resource (Takahashi et al. 2007, Takahashi and Yamanaka 2006, Yu et al. 2007). Moreover, the ability to remove the inducing genes (Woltjen et al. 2009), the addition of only one or two genes in certain cell types (Kim et al. 2009) and generation of iPSCs by chemical induction (Li et al. 2009, Lin et al. 2009) has of course highlighted these as an alternative potential source of therapeutic stem cells which are not embryonic in origin. As such, they avoid the ethical controversy; the technology even (though not relevant in the context of use in a bioartificial liver) offers the possibility of generating liver cells with the same immunogenetic background as the host (Ghodsizadeh et al. 2010); and it is already clear that the maturation profiles of these cells, and the techniques required to induce hepatocytic function, are very similar to those pioneered with ESCs (Hay et al. 2008, Sullivan et al. 2010).

A large amount of literature is accumulated with mouse ESCs, which mainly lies outside the scope of this chapter, but the subsequent development of protocols for human cells are mentioned. The key in principle for both ESCs and IPSCs is the recapitulation of the events of embryogenesis—the maturation and specification of the endoderm in response to adjacent mesodermal tissues, hepatic induction and liver programming. The concentrations and indeed gradients of different growth factors during these stages are critical, with notably a reducing gradient of mesodermal factors such as Fibroblast Growth Factors, and interactions between the latter and bone morphogenic proteins, being necessary. Replicating these changes to initiate hepatocyte differentiation *in vitro* has resulted in description of a variety of "recipes" to this end.

The last few years have seen an increasing number of laboratories progressively refining approaches to ESC and then IPSC differentiation, modifying a series of papers in the early 2000s that had modulated differentiation of mouse ESCs (Lavon et al. 2004). They characterized expression level liver-enriched genes in undifferentiated and differentiated hESCs showing a subset of foetal hepatic enriched genes becoming expressed upon differentiation into embryoid bodies. Using a hepatocyte-specific promoter for GFP they sorted differentiated cells and expanded them for some weeks, but did not elicit specific adult hepatocyte functions

Hay et al. (2007) reported a multistage procedure by which hESCs can be directly differentiated to hepatocyte-like cells without embryoid body formation with activin A and sodium butyrate prior to further differentiation to hepatocytes with dimethyl sulphoxide, followed by maturation with hepatocyte growth factor and oncostatin M. (Activin enhances expression of a panel of endodermal genes such as SOX 17 and Fox A2). The hESC-derived hepatocyte-like cells (HLCs) exhibited characteristic hepatocyte morphology, expressed alpha-fetoprotein, albumin, and hepatocyte nuclear factor 4alpha, and possessed hepatocyte-specific activities, such as p450 metabolism (Cyp3A4, Cyp1A2, Cyp2C9/19 and Cyp2D6), albumin production, glycogen storage, and uptake and excretion of indocyanine green. Agarwal et al. (2008) generated high-purity (70%) hepatocyte cultures from hESCs using sequentially Activin A, FGF-4 and HGF, and oncostatin. Cell cultures progressively expressed increasing numbers of hepatic lineage markers, including GATA4, HNF4α, α-fetoprotein, CD26, albumin, α-1-antitrypsin, Cyp7A1, and Cyp3A4., and functionally demonstrated glycogen storage, indocyanine green uptake and release, and albumin secretion. In a mouse model of acute liver injury, the hESC-derived definitive endoderm differentiated into hepatocytes and repopulated the damaged liver. These papers provided a blueprint on which further refinement of human hepatocyte differentiation protocols have been developed, although it itself grew from many observations on mouse ESCs, summarized by Llavon and Benevisty (2005).

Subsequently Hay et al. (2008) developed the concept of recapitulation of *in vivo* liver differentiation by harnessing the role that Wnt signalling has been shown to play coordinating development, showing its differential expression at critical stages of human liver development *in vivo,* and demonstrated that it could play a similar role *in vitro.*, with Wnt3a facilitating clonal plating of hESCs exhibiting functional hepatic differentiation. This was followed by Basma et al. (Basma et al. 2009) who generated hepatocyte-like cells from human ESCs by sequential culture in fibroblast growth factor 2 and human activin-A, hepatocyte growth factor, and dexamethasone, then isolating functional hepatocytes by sorting for surface asialoglycoprotein-receptor expression. Most recently, Touboul et al. (2010) differentiated hESCs into a homogenous population of endoderm cells using a combination of activin, fibroblast growth factor 2 and bone morphogenetic protein 4, together with phosphoinositide 3-kinase inhibition. The endoderm cells were then induced to differentiate further into hepatic progenitors using fibroblast growth factor 10, retinoic acid, and an inhibitor of activin/nodal receptor. After maturation, these cells expressed the asialoglycoprotein receptor, tyrosine aminotransferase, alpha1-antitrypsin, Cyp7A1, and hepatic transcription factors such as hepatocyte nuclear factors 4alpha and 6. Again, cells generated under these conditions exhibited hepatic functions *in vitro*, including glycogen storage, cytochrome activity, and low-density lipoprotein uptake. After transduction with a green fluorescent protein-expressing lentiviral vector and transplantation into immunodeficient uPA transgenic mice, differentiated cells engrafted into the liver, grew, and expressed human albumin and alpha1-antitrypsin as well as green fluorescent protein for at least 8 wk. More recent endeavours with hESCs have addressed standardization of ESC tissue culture, notably by employment of serum-free media (Hannoun et al. 2010), rather than mouse embryonic fibroblast conditioned media and synthetic extracellular supports (Hay et al. 2011). With two serum free media hESCs were maintained in a pluripotent state, for over 30 passages, and hESCs maintained in these media generate equivalent levels of human hepatic endoderm as observed with CM, but demonstrated superior function *in vitro*. Moreover the use of wholly synthetic substrates in 2D and 3D promoted hepatocyte metabolic activity and drug inducible function *in vitro* (Hay et al. 2011).

Song et al. (2009) were the first to describe the expression of liver cell markers and liver-related functions from human iPSC-derived cells, comparing their properties with that of differentiated human ES cells and primary human hepatocytes. Approximately 60% of the differentiated human IPSCs at day 7 expressed hepatic markers alpha fetoprotein and Alb. By day 21differentiated cells exhibited albumin secretion, glycogen synthesis, urea production and inducible cytochrome P450 activity, at levels comparable to those of human ES cell-derived hepatic cells.

In 2010 further independent laboratories reported complete derivation of iPSCs into endodermal-derived hepatocytes (Sullivan et al. 2010, Si-Tayeb et al. 2010) with hepatocytic markers and expression of hepatocyte-specific function. These studies demonstrated that iPSC lines representing both male and female sexes and different ethnic origins could be differentiated to hepatocyte-like cells at efficiencies of between 70–90%. The iPSC-derived HE exhibited hepatic morphology and expressed the hepatic markers albumin and E-cadherin, alpha-fetoprotein, hepatocyte nuclear factor-4a, and produced and secreted the plasma proteins, fibrinogen, fibronectin, transthyretin, and alpha-fetoprotein. Both studies showed that IPSC derived hepatocytes demonstrated CYP1A2, and notably CYP3A4 function (responsible for ~70% of xenobiotic metabolism). In addition, Si Tayeb et al. (2010) eliminated the use of serum and feeder cell layers dispensing thus with undefined reagents which characterize many protocols previously applied to mouse and human ESC maturation protocols. A highly laudatory editorial, commented that this work confirmed that and IPSCs could be used in every respect to replace ESCs for liver regeneration and for studying pathogens as the cause of hepatocyte and liver dysfunction.

Clearly this claim needs critical assessment. Chin et al. (2009) reported in contrast that despite IPSC and hESC similarities, there was a distinguishing recurrent gene expression signature extending to miRNA expression, probably reflecting preferential usage of promoter sequences. Jozefczuk et al. (2011) compared hepatocyte-like cells (HLCs) derived from IPSCs cells to those from hESCs at the transcriptional level. Two independent protocols for deriving HLCs from hESCs and iPSCs were used, and analysis included immunocytochemistry, real-time (RT)-polymerase chain reaction, and *in vitro* functional assays and comparative microarray-based gene expression profiling. Whilst HLCs derived from hESCs and human iPSCs showed significant functional similarities, and similar expression of genes important for liver physiology and common pathways, there were specific differences between the 2 cell types. For example among the cytochrome P450 gene family, CYP19A1, CYP1A1, and CYP11A1 were enriched in hESC-derived HLCs, and CYP46A1 and CYP26A1 in iPSC-derived HLCs. This study of course used specific differentiation protocols and it is scarcely surprising that the outcomes do differ in certain ways. The principles used for IPSC induction, particularly the use or not of viral gene transfer, are likely to be important future issues in defining the "final product" of an IPSC derived hepatocyte.

Furthermore of course the individual donor source of both the ESC line and the IPSCs will be reflected in the final cell derived. This has been most dramatically exemplified by the series of papers which have used disease specific fibroblasts to produce cell-lines modelling disease; but in the future clearly there is the potential for unappreciated polymorphisms in cell derived from different human hESC and iPSCs to affect final performance.

There have been studies of alternative sources of stem cells—in particular those that are relatively easily available for study. Seo et al. (2005) examined human adipose tissue-derived stromal cells —hADSC—a mesenchymal stem cell—and demonstrated that treatment with HGF and OSM and DMSO could initiate differentiation into cells morphologically similar to hepatocytes, expressing albumin and AFP at RNA and immunocytochemical level, taking up LDL and producing urea. However levels were less functional than HepG2 monolayers. Extra-embryonic yolk sac tissue is also a potential source—Chien et al. (2006) derived multipotent cells from human term placenta showing that in a Hepatocyte differentiation medium including HGF and FGF-4, albumin and the hepatocyte associated cytokeratin CK18 were expressed and RNA levels CYP3A4 were induced in response to rifampicin treatment.

Cultured in what form? Maintained in what kind of extracorporeal chamber (bioreactor)?

As we have discussed elsewhere, the links between an *in vitro* demonstration of functional activity, and the utilization of a cell in a bioartificial liver, is a great deal more complex than simple extrapolation from the small scale demonstration of expression of activity. The architecture that cells within a bioreactor will exhibit is of key importance, involving not only a choice of supporting matrices—which must be compatible with use in an extracorporeal system, but also the microscopic structure imposed on the cells thereby. Whilst only little explored thus far with cell lines differentiated from stem cells, Miki et al. (2005) used a four-compartment 3D perfusion culture (the Berlin reactor) to induce aid induction of hepatic differentiation in hESC. Undifferentiated hESC were inoculated into hollow fibre-based 3D perfusion bioreactors with integral oxygenation. Hepatic differentiation was induced with a multistep growth factor cocktail protocol. Parallel controls were operated under equal perfusion conditions without the growth factor supplementations to allow for spontaneous differentiation, as well as in conventional 2D static conditions using growth factors. Significantly upregulated hepatic gene expression was observed in the hepatic differentiation 3D culture group with ammonia metabolism activity albumin production, drug-induced Cytochrome P450 gene expression, was observed in the 3D directed differentiation culture. Although, perhaps disappointingly, asialoglycoprotein receptor expression a marker of mature hepatocyte generation was found on only up to 30% of the cells in the 3D system with directed hepatic differentiation.

Perhaps more concerningly—and certainly a major hurdle currently in practical terms, there is a need for a quantum leap (downwards) in the cost of generation of hepatocytes from these stem cell sources. Faced with the requirement of generating between 10^{10} and 10^{11} cells, the costs notably of media and growth factor supplements are truly formidable. One unofficial estimate of the cost of generation between 10^{6} and 10^{7} cardiomyocytes from hECSs produced a cost of ~$1000; the possibility currently of entering clinical trials with a machine of commercial viability seem small, and cost savings on a major scale will need to be achieved.

One may conclude that the next few years will see better protocols —hopefully cheaper—for generating hepatocytes with a near-normal *ex vivo* functional repertoire. It seems that the testing-bed for such devices is likely to be the pharmaceutical industry, where high throughput screens reflecting normal hepatocyte metabolism would be invaluable—and a huge advance on current approaches using rodent cells or human primary cells recovering from cryopreservation injury. The extracorporeal liver support systems will in the short term be developed using either tumour cell lines or immortalized cells, and will wait on a robust and economic supply of progenitor derived cells.

References

Agarwal S, KL Holton and R Lanza. 2008. Efficient differentiation of functional hepatocytes from human embryonic stem cells. Stem Cells. 26(5): 1117–27.

Araki N, STsuruoka, N Wang et al. 2008. Human CYP3A4-introduced HepG2 cells: *in vitro* screening system of new chemicals for the evaluation of CYP3A4-inhibiting activity. Xenobiotica. 38: 1355–64.

Banares R, F Neven, FS Larsen et al. 2010. Extracorporeal liver support with the MARS system in patients with acute-on-chronic liver failure. J Hepatol. 52 : S459.

Basma H, A Soto-Gutiérrez, GR Yannama et al. 2009. Differentiation and transplantation of human embryonic stem cell-derived hepatocytes. Gastroenterology. 136(3): 990–9.

Cederbaum SD, H Yu, WW Grody et al. 2004. Arginase I and II: do their functions overlap? Molecular Genetics and Metabolism. S38–S44.

Chien CC, BL Yen, FK Lee and et al. 2006. *In vitro* differentiation of human placenta-derived multipotent cells into hepatocyte-like cells. Stem Cells 24(7): 1759–68.

Chin MH, MJ Mason, W Xie et al. 2009. l. Induced pluripotent stem cells and embryonic stem cells are distinguished by gene expression signatures. Stem Cell. 5(1): 111–23.

Christensen E, A Bremmelgaard, M Bahnsen et al. 1984. Prediction of fatality in fulminant hepatic failure. Scand J Gastroenterol. 19(1): 90–6.

Demetriou AA, RS Brown Jr, RW Busuttil et al. 2004. Prospective, randomized, multicenter, controlled trial of a bioartificial liver in treating acute liver failure. Ann Surg. 239(5): 660–7.

Deurholt T, ten L Bloemendaal, AA Chhatta et al. 2006. *In vitro* functionality of human fetal liver cells and clonal derivatives under proliferative conditions. Cell Transplant. 15: 811–22.

Donato MT, N Jimenez, A Serralta et al. 2007. Effects of steatosis on drug-metabolizing capability of primary human hepatocytes. Toxicol *In Vitro*. 21: 271–6.

Dorrell C and M Grompe. 2005. Liver repair by intra- and extrahepatic progenitors. Stem Cell Rev. 1(1): 61–4.

Fontana RJ, KS Lown, MF Paine et al. 1999. Effects of a chargrilled meat diet on expression of CYP3A, CYP1A, and P-glycoprotein levels in healthy volunteers. Gastroenterology. 117: 89–98.

Ghodsizadeh A, A Taei, M Totonchi, and 2010. Generation of liver disease-specific induced pluripotent stem cells along with efficient differentiation to functional hepatocyte-like cells. Stem Cell Rev. 6(4): 622–32.

Hannoun Z, J Fletcher, S Greenhough et al. 2010. The Comparison between Conditioned Media and Serum-Free Media in Human Embryonic Stem Cell Culture and Differentiation Cellular Reprogramming. 12(2): 133–140.

Hariparsad N, BA Carr, R Evers et al. 2008. Comparison of immortalized Fa2N-4 cells and human hepatocytes as *in vitro* models for cytochrome P450 induction. Drug Metab Dispos. 36: 1046–55.

Harmsen S, AS Koster, JH Beijnen et al. 2008. Comparison of two immortalized human cell lines to study nuclear receptor-mediated CYP3A4 induction. Drug Metab Dispos. 36: 1166–71.

Hay DC, D Zhao, A, Ross et al. 2007. Direct differentiation of human embryonic stem cells to hepatocyte-like cells exhibiting functional activities. Cloning Stem Cells. 9(1): 51–62.

Hay DC, J Fletcher, C Payne et al. 2008. Highly efficient differentiation of hESCs to functional hepatic endoderm requires Activin A and Wnt3a signaling. Proc Natl Acad Sci USA. 105(34): 12301–6.

Hay DC, S Pernagallo, JJ Diaz-Mochon et al. 2011. Unbiased Screening of Polymer Libraries to Define Novel Substrates for Functional Hepatocytes with Inducible Drug Metabolism. Stem Cell Research. 6: 92–101.

Horslen SP, JM Hammel, LW Fristoe et al. 2000. Extracorporeal liver perfusion using human and pig livers for acute liver failure. Transplantation. 70(10): 1472–8.

Jozefczuk J, A Prigione, L Chavez et al. 2011. Comparative Analysis of Human Embryonic Stem Cell and Induced Pluripotent Stem Cell-Derived Hepatocyte-Like Cells Reveals Current Drawbacks and Possible Strategies for Improved Differentiation Stem Cells and Development.

Kanebratt KP and TB Andersson. 2008. Evaluation of HepaRG cells as an *in vitro* model for human drug metabolism studies. Drug Metab Dispos. 36: 1444–52.

Kenny JR, L Chen, DF McGinnity et al. 2008. Efficient assessment of the utility of immortalized Fa2N-4 cells for cytochrome P450 (CYP) induction studies using multiplex quantitative reverse transcriptase-polymerase chain reaction (qRT-PCR) and substrate cassette methodologies. Xenobiotica. 38: 1500–17.

Kim JB, B Greber, MJ Araúzo-Bravo et al. 2009. Direct reprogramming of human neural stem cells by OCT4. Nature. 461: 649–643.
Klebanoff G, D Hollander, RT Osteen et al. 1973. Human-baboon cross circulation in hepatic failure: report of a successful case. Mil Med. 138(3): 156–9.
Kundra A, A Jain, A Banga et al. 2005. Evaluation of plasma ammonia levels in patients with acute liver failure and chronic liver disease and its correlation with the severity of hepatic encephalopathy and clinical features of raised intracranial tension. Clinical Biochemistry. 38: 696–699.
Kuwahara R, AV Kofman, CS, Landis et al. 2008.The hepatic stem cell niche: identification by label-retaining cell assay. Hepatology. 47(6): 1994–2002.
Lavon N and N Benvenisty. 2005. Study of hepatocyte differentiation using embryonic stem cells. J Cell Biochem. 96(6): 1193–202.
Lavon N, O Yanuka and N Benvenisty. 2004. Differentiation and isolation of hepatic-like cells from human embryonic stem cells. Differentiation. 72(5): 230–8.
Li W, H Zhou, R Abujarour et al. 2009. Generation of human induced pluripotent stem cells in the absence of exogenous Sox2. Stem Cells doi: 10.1002/stem.240.
Lim PL, W Tan, and C Latchoumycandane et al. 2007. Molecular and functional characterization of drug-metabolizing enzymes and transporter expression in the novel spontaneously immortalized human hepatocyte line HC-04. Toxicol *In Vitro*. 21: 1390–401.
Lin T, R Ambasudhan, X Yuan et al. 2009. A chemical platform for improved induction of human iPSCs. Nat Methods. 6: 805–808.
Maruyama M, T Matsunaga, E Harada et al. 2007. Comparison of basal gene expression and induction of CYP3As in HepG2 and human fetal liver cells. Biol Pharm Bull. 30: 2091–7.
Matsunaga T, M Maruyama, E Harada et al. 2004. Expression and induction of CYP3As in human fetal hepatocytes. Biochem Biophys Res Commun. 318: 428–34.
Matsushita T, I Ikai, R Nishitai et al. 2003. Suppressed complement activation in human decay accelerating factor transgenic porcine liver cross-circulated with nonhuman primates. Transplantation. 75(11): 1807–12.
Mavri-Damelin D, LH Damelin, S Eaton et al. 2008. Cells for bioartificial liver devices: the human hepatoma-derived cell line C3A produces urea but does not detoxify ammonia. Biotechnol Bioeng. 99: 644–51.
Mavri-Damelin D, S Eaton, LH Damelin et al. 2007. Ornithine transcarbamylase and arginase I deficiency are responsible for diminished urea cycle function in the human hepatoblastoma cell line HepG2. Int J Biochem Cell Biol. 39: 555–64.
Miki T, A Ring, and J Gerlach. 2011. Hepatic Differentiation of Human Embryonic Stem Cells is Promoted by Three-Dimensional Dynamic Perfusion Culture Conditions Tissue Engineering: Part C. 17; 5: 11.
Poyck PP, ACvan Wijk, TV van der H(et al. et al. 2008. Evaluation of a new immortalized human fetal liver cell line (cBAL111) for application in bioartificial liver. J Hepatol. 48: 266–75.
Poyck PP, R oekstra, AC van Wijk et al. 2007. Functional and morphological comparison of three primary liver cell types cultured in the AMC bioartificial liver. Liver Transpl. 13: 589–98.
Rifai K, A Kribben, G, Gerken et al. 2010. Extracorporeal support by Fractionated Plasma separation and Adsorption in acute-on-chronic liver failure. J Hepatol. 52: S3.
Rodriguez-Antona C, MT Donato, A Boobis et al. 2002. Cytochrome P450 expression in human hepatocytes and hepatoma cell lines: molecular mechanisms that determine lower expression in cultured cells. Xenobiotica. 32: 505–20.
Seo MJ, SY Suh, YC Bae et al. 2005. Differentiation of human adipose stromal cells into hepatic lineage *in vitro* and *in vivo*. Biochem Biophys Res Commun. 328(1): 258–64.
Si-Tayeb K,FK Noto, M Nagaoka et al. 2010. Highly efficient generation of human hepatocyte-like cells from induced pluripotent stem cells. Hepatology. 51: 297–305.
Smalley M, K Leiper, R Tootle et al. 2001. Immortalization of human hepatocytes by temperature-sensitive SV40 large-T antigen. *In Vitro* Cell Dev Biol Anim. 37(3): 166–8.

Song Z, J Cai, Y Liu et al. 2009. Efficient generation of hepatocyte-like cells from human induced pluripotent stem cells. *Cell Res.* 19(11): 1233–42.

Starzl TE, CW Putnam, CG Groth et al. 1975. Alopecia, ascites, and incomplete regeneration after 85 to 90 per cent liver resection. Am J Surg. 129: 587–90.

Sullivan GJ,,DC Hay, IH Park and et al. 2010. Generation of functional human hepatic endoderm from human induced pluripotent stem cells. Hepatology. 51: 329–335.

Takahashi K, K Tanabe, M Ohnuki et al. 2007. Induction of pluripotent stem cells from adult human fibroblasts by defined factors. Cell. 131: 861–872.

Takahashi Kand S Yamanaka. 2006. Induction of pluripotent stem cells from mouse embryonic and adult fibroblast cultures by defined factors. Cell. 126: 663–676.

Touboul T, NR Hannan, S Corbineaua et al. 2010. Generation of functional hepatocytes from human embryonic stem cells under chemically defined conditions that recapitulate liver development. Hepatology. 51(5): 1754–65.

Wege H, HT Le, MS Chui et al. 2003. Telomerase reconstitution immortalizes human fetal hepatocytes without disrupting their differentiation potential. Gastroenterology. 124(2): 432–44.

Wilkening S, F Stahl and A Bader. 2003. Comparison of primary human hepatocytes and hepatoma cell line HepG2 with regard to their biotransformation properties. Drug Metab Dispos. 31: 1035–42.

Woltjen K, IP Michael, P Mohseni et al. 2009. piggyBac transposition reprograms fibroblasts to induced pluripotent stem cells. Nature 458: 766–770.

Yan Z and GW Caldwell. 2001. Metabolism profiling, and cytochrome P450 inhibition & induction in drug discovery. Curr Top Med Chem. 1: 403–25.

Yang Q, F Liu, XP Pan et al. 2010. Fluidized-bed bioartificial liver assist devices (BLADs) based on microencapsulated primary porcine hepatocytes have risk of porcine endogenous retroviruses transmission. Hepatol Int. 4(4): 757–61.

Yoshitomi S, K Ikemoto, J Takahashi et al. 2001. Establishment of the transformants expressing human cytochrome P450 subtypes in HepG2, and their applications on drug metabolism and toxicology. Toxicology *in Vitro*. 15: 245–56.

Yu J, MAVodyanik, K Smuga-Otto et al. 2007. Induced pluripotent stem cell lines derived from human somatic cells. Science. 318: 1917–1920

9

Stem Cell Technology and Cell Based Therapies

Nicholas James Davies[1,a,*] and
Philip Noel Newsome[1,b]

Introduction

The liver is the major metabolic organ of the body performing a large number of endocrine and exocrine functions. It also possesses a unique capacity to regenerate as is demonstrated by the complete regeneration of the liver following two thirds partial hepatectomy.

Currently, orthotopic liver transplantation (OLT) is the only cure for end-stage liver disease, as well as a number of inherited metabolic disorders (Shneider 2002, Fisher and Storm 2006). The procedure is invasive, associated with relatively high mortality and morbidity as well as high cost. Furthermore, there is much greater demand for livers than can be met by donors, which has led to the need for the development of alternative therapies to treat liver disease. This chapter will discuss these therapies and highlight the potential role that embryonic stem cells and inducible pluripotent stem cells (iPSCs) may play.

Cellular transplantation

Hepatocyte transplantation may be a potential alternative to OLT or may provide a bridge for patients on liver transplant waiting lists. It could be

[1]Centre for Liver Research and NIHR Biomedical Research Unit, University of Birmingham, Birmingham, U.K.
[a]Email: n.j.davies@bham.ac.uk
[b]Email: p.n.newsome@bham.ac.uk
*Corresponding author

particularly relevant for those patients with inborn errors in metabolism such as Crigler–Najjar syndrome , urea cycle disorders and glycogen storage disorders, where the cause is a genetic defect affecting hepatocyte function. In these cases, unlike most adult disorders, the architecture of the liver is intact (Grompe 2006) and there is likely to be less need to fully repopulate the liver with donor hepatocytes (Grompe 2006, Dhawan et al. 2006, Enns and Millan 2008).

Transplanted cells have to run the gauntlet from the site of infusion, whether that be the spleen, as is commonly used in smaller mammals, or the portal vein, to the hepatic sinusoids and across the space of Disse to engraft before they are able to integrate into the parenchyma (Fig. 9.1). In addition to correct homing to damaged tissue, donor hepatocytes need to

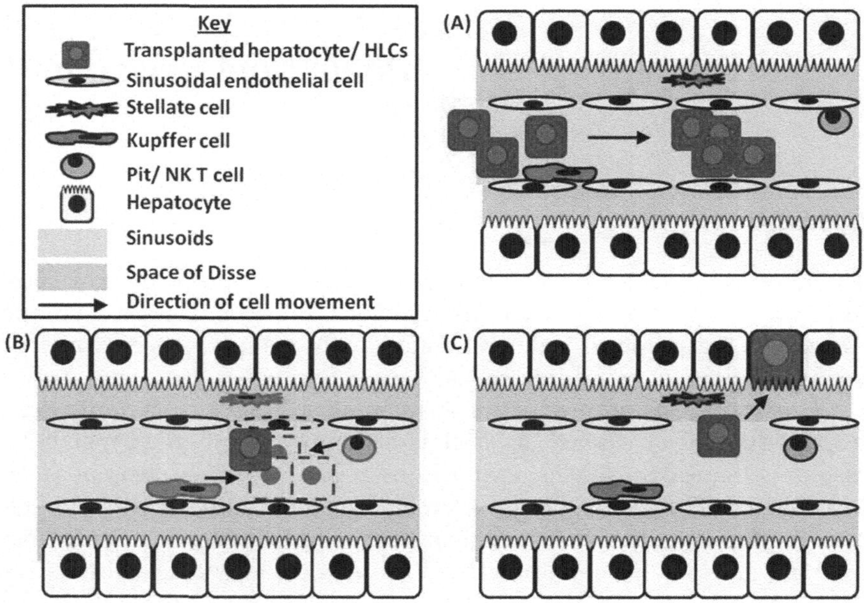

Figure 9.1. The route to engraftment for transplanted hepatocytes and hepatocyte-like cells. Schematic representation of the fate of transplanted hepatocytes and HLCs upon entering the liver sinusoids.

A. Transplanted hepatocytes/hepatocyte-like cells enter the sinusoids where they become lodged.
B. The majority of hepatocytes undergo anoikis and/or are removed as a result of clearance by Kupffer cells and Pit/NK T cells. Stellate cells become activated and release factors which may contribute to the disruption of sinusoidal endothelial barrier.
C. Following disruption of sinusiodal endothelial barrier, hepatocytes cross the Space of Disse and integrate into the parenchyma.

Color image of this figure appears in the color plate section at the end of the book.

be able to successfully evade the immune system both in the periphery and within the liver. Kuppfer cells are liver-resident macrophages which are rapidly recruited to the sinusoids following cellular transplantation (Joseph et al. 2002, Wilhelm et al. 2004) where they are thought to phagocytose transplanted cells (Joseph et al. 2002). Inhibition of Kuppfer cell function using gadolinium chloride has been demonstrated to increase hepatocyte engraftment (Joseph et al. 2002). Furthermore, depletion of natural killer cells and neutrophils has also been demonstrated to enhance cellular engraftment in the liver (Kawahara et al. 2010b, Krohn et al. 2009), suggesting a role for these cells in the eradication of transplanted cells. Post immune evasion, transplanted cells need to be able to adhere to hepatic sinusoidal endothelial cells and/or the underlying ECM to prevent detachment-induced apoptosis or anoikis (Smets et al. 2002). Following attachment, transplanted hepatocytes need to overcome the barrier posed by the sinusoidal endothelium and transmigrate to and engraft in the liver parenchyma (Fig. 9.1). In experiments using rats it has been demonstrated that following cellular transplantation the sinusoidal endothelium becomes disrupted, possibly as a result of growth factors, such as VEGF which are released by activated stellate cells and hepatocytes (Lee et al. 2007, Gupta et al. 1999). The resultant loss of integrity of the hepatic sinusoidal endothelium permits transplanted hepatocytes to cross the space of Disse and integrate into the liver parenchyma (Fig. 9.1).

Hepatocyte transplantation has been demonstrated to be successful in rodents, as well as larger mammals including non-human primates (Matas et al. 1976, Ponder et al. 1991, Groth et al. 1977, Kocken et al. 1996, Selden et al. 1995, Sommer et al. 1980, Wiederkehr et al. 1990, Dagher et al. 2009), with a clinical improvement in several animal models of human diseases. Both allogeneic and xenogeneic hepatocytes have been shown to engraft and function in animal models (Rhim et al. 1995, Makowka et al. 1980, Nagata et al. 2003). These include the transplantation of human hepatocytes, which have been shown to be capable of reconstituting the livers in three transgenic murine models (Mercer et al. 2001, Washburn et al. 2011, Azuma et al. 2007, Bissig et al. 2007, Kawahara et al. 2010a).

Hepatocyte transplantation

As only a small number of transplanted hepatocytes successfully engraft into the liver parenchyma, it is beneficial that once they have engrafted they are able to proliferate with a survival advantage over host hepatocytes (Grompe 2006).

Experimental models to improve liver repopulation

The $FAH^{Exon\Delta 5}$/tyrosinaemia, alb-uPA and AFC8 mouse models each give the donor hepatocytes a proliferative advantage over the host hepatocytes (Washburn et al. 2011, Overturf et al. 1999, Song et al. 2009, Rhim et al. 1994). This occurs either through the build up of hepatotoxic by-products of tyrosinaemia catabolism in the case of the $FAH^{Exon\Delta 5}$ mouse (Grompe et al. 1993, Grompe et al. 1995), or in the case of the alb-uPA and AFC8 mouse models, the hepatocyte-specific expression of proteins which induce hepatocyte cell death (Washburn et al. 2011, Song et al. 2009, Rhim et al. 1994). This highlights an important consideration for cellular therapy, the need for a preparative regimen to give the transplanted cells the best chance of engrafting and proliferating in the host liver.

In 1998 Laconi et al. reported that the plant alkaloid Retrorsine could be used in conjunction with two thirds partial hepatectomy to give near total replacement of the hepatocyte mass with transplanted cells (Laconi et al. 1998). This protocol has been successfully used by Basma et al. to transplant human ES-derived hepatocytes into mice. Although Retrorsine and other plant alkaloids, such as Monocrotaline (Joseph et al. 2006) increase donor hepatocyte repopulation by inhibiting host hepatocyte proliferation and, as such, are permissive to improving engraftment of donor cells, they are carcinogenic (Fu et al. 2004). Therefore, it is preferable to identify other means of providing donor hepatocytes with a selective growth advantage.

The identification and manipulation of molecules associated with hepatocyte proliferation is one potential means to give donor hepatocytes a proliferative advantage over host cells. This was elegantly demonstrated by Karnezis et al who demonstrated that hepatocytes from $p27^{Kip1}$ knockout mice displayed better repopulation and engraftment into $FAH^{Exon\Delta 5}$ mice compared to hepatocytes derived from wild-type mice (Karnezis et al. 2001). $p27^{kip1}$ negatively regulates cell proliferation through interactions with various cyclin/Cdk complexes (Sherr and Roberts 1995, Sherr and Roberts 1999) and knockout mice display multiorgan hyperplasia due to increased cell numbers (Fero et al. 1996, Nakayama et al. 1996, Kiyokawa et al. 1996). The authors demonstrated that the *in vitro* response of $p27^{Kip1-/-}$ hepatocytes to epidermal growth factor stimulation was both greater and quicker than wild type hepatocytes which is likely to provide a significant proliferative advantage to transplanted $p27^{Kip1-/-}$ hepatocytes over host liver cells in response to an exogenously applied mitogen (Karnezis et al. 2001).

Allogeneic hepatocytes have been demonstrated to be susceptible to apoptosis as a result of Fas-Fas ligand (FasL) interactions (Kawahara et al.

1997). Transplantation of Fas-resistant hepatocytes constitutively expressing anti-apoptotic genes such as BCL_2 or BCL_{XL}, have shown improved engraftment compared to non-transgenic hepatocytes (Wang et al. 2004, Mitchell et al. 2002, Mignon et al. 1998, Mitchell et al. 2000). Furthermore, these modifications will provide the donor cells with a survival advantage over host hepatocytes when treated with Fas-activating antibodies thereby allowing for enhanced levels of repopulation (Mignon et al. 1998, Mitchell et al. 2000).

Targetted hepatic irradiation has been used clinically to treat hepatocellular carcinoma and tumour metastases to the liver, but the high levels of irradiation required can lead to potentially lethal liver damage (Fajardo and Colby 1980, Dawson and Ten Haken 2005). However, sub-lethal levels of irradiation have been used in rodent models to augment hepatocyte transplantation (Guha et al. 2001, Guha et al. 1999). Radiation induces DNA damage which prevents the ability of hepatocytes to proliferate as a result of a mitogen causing them to undergo apoptosis. Therefore, donor hepatocytes transplanted into mice that have received hepatic irradiation have a proliferative advantage over host hepatocytes, permitting high levels of repopulation by donor cells in response to a proliferative stimulus provided either by T_3 treatment or by injurious agents such as carbon tetrachloride or adenoviral-delivery of FasL (Guha et al. 2001, Takahashi et al. 2003). Furthermore, it is possible to target individual lobes by shielding the rest of the liver with lead, which may be of particular relevance for the treatment of metabolic disorders where whole liver repopulation is not necessary.

Using intravital microscopy, Gupta et al. (1999) demonstrated that transplanted hepatocytes adhere to the hepatic sinusoidal lumen for up to 16 hr before crossing this endothelial barrier and integrating into the parenchyma. It is essential that transplanted hepatocytes adhere otherwise they will undergo anchorage-dependent apoptosis or anoikis (Smets et al. 2002). Enhanced engraftment of hepatocytes has been demonstrated by posting ECM components onto sinusoidal cells prior to transplantation (Kumaran et al. 2005).

Disruption of the sinusoidal endothelium using Cyclophosphamide or Monocrotaline was demonstrated to augment engraftment of hepatocytes (Joseph et al. 2006, Malhi et al. 2002). Gupta et al. demonstrated that combinations of Phenytoin or Rifampicin potentiated the effect of Monocrotaline on endothelium via activation of cytochrome P450 enzymes (Wu et al. 2008). Indeed, hepatic irradiation may augment hepatocyte engraftment by transient disruption of hepatic sinusoidal cells (Yamanouchi et al. 2009).

Clinical hepatocyte transplantation

The first reported clinical hepatocyte transplants were performed by Grossman and colleagues to treat five patients with familial hypercholesterolemia. This protocol used gene therapy to correct the genetic defect in the patients' own hepatocytes (Grossman et al. 1994, Grossman et al. 1995) and resulted in a greater than 20% reduction in the cholesterol of three out of five patients. However, less than 5% of transgene expression was observed after 4 mon (Raper et al. 1996). In 1998 Fox et al. reported the world's first autologous hepatocyte transplantations to treat two patients suffering from Criglar-Najjar Syndrome (Fox et al. 1998). Whilst this led to an improvement in their clinical outlook, the improvement was only transient and both patients later received OLT. Hepatocyte transplantation has been used to treat a wide range of patients suffering with metabolic disorders where improvements in clinical outlook have been reported (reviewed in Fisher and Strom 2006). However, these improvements have only been transient due to either loss of donor hepatocyte function or viability.

Hepatocyte transplantation has also been used to treat patients suffering from both acute and chronic liver disease. In these cohorts of patients, the hepatocyte transplants have often served as a bridge to OLT. However, there have been several cases, particularly with patients with acute liver injury, where a full recovery has been made without the need for OLT (reviewed in Fisher and Strom 2006). In these scenarios, it may be that the donor hepatocytes have served to permit host hepatocyte regeneration by alleviating their functional burden. Aside from these few exceptions, to date hepatocyte transplantation has primarily served as a bridge to OLT as donor hepatocytes have not survived or maintained their function for prolonged periods of time (Fisher and Strom 2006).

Alternatives to primary human hepatocytes

Although hepatocyte transplantation has the advantage that it is a less invasive procedure than OLT and it may be possible to use cells from livers that would be rejected for OLT, there is still the problem that demand will still outstrip supply. This is the case even if multiple recipients are treated with cells isolated from a single liver. Hepatocytes isolated from human cadavers may provide another source of cells for transplantation. Indeed, cells isolated from the cadaveric livers of mice and humans have been demonstrated to engraft and function in the $FAH^{Exon\Delta5}$ mouse (Erker et al. 2010). Furthermore, in competitive repopulation studies, the hepatcoytes isolated from cadaveric livers demonstrated similar levels of repopulation compared to cells isolated from live mice, suggesting that they still possess the ability to engraft and proliferate. However, other sources of cells need to be identified to meet the demand of transplant waiting lists.

Transformed hepatocytes

Serial transplant experiments using the $FAH^{Exon\Delta5}$ mouse model demonstrated that hepatocytes potentially have almost unlimited proliferative capacity (Overturf et al. 1999). However, the proliferative capacity for isolated hepatocytes is limited *in vitro*, even when the cells are stimulated using mitogens such as epidermal growth factor, transforming growth factor α or hepatocyte growth factor. Furthermore, both viability and function of primary hepatocytes is short-lived *in vitro*. To overcome the limited proliferative capacity of hepatocytes *in vitro* several groups have used the Simian Virus (SV40) large T antigen to transform primary hepatocytes (Kobayashi et al. 2000, Kim et al. 2000, Schumacher et al. 1996, Allain et al. 2002) to generate functional hepatocytes which can proliferate indefinitely at permissive temperatures. However, this transformation required the integration of viral genes into the host genome and is therefore a likely tumour risk.

Immortalized hepatocytes have been demonstrated to engraft and function in animal models, including acute liver failure and the Gunn rat, where the cells were found to improve survival and increase serum levels of conjugated bilirubin, respectively (Kobayashi et al. 2000, Nakamura et al. 1997, Tada et al. 1998). However, SV40 large T antigen transformed hepatocytes have been reported to undergo a proliferative and functional crisis with prolonged culture (Kim et al. 2000). One reason for this crisis in proliferation is likely to be due to shortening in telomere length (Delgado et al. 2005). Pan and colleagues have succeeded in generating immortalized porcine hepatocytes using co-transduction of SV40 large T antigen and hTERT (Pan et al. 2010). Whilst this approach may have uses for extracorporeal devices, such as bio-artificial livers, their potential for cellular transplantation is severely limited given their inability to contribute to normal tissue homeostasis.

Bone marrow-derived cells

The bone marrow has several types of resident stem cell including haematopoietic and mesenchymal stem cells. Haematopoietic stem cells (HSCs) have been demonstrated to be remarkably plastic cells, with the ability to differentiate into cells of non-haematopoietic lineages including hepatocytes, myocytes and neural cells (Ferrari et al. 1998, Brazelton et al. 2000, Mezey et al. 2000, Newsome et al. 2003, Ishikawa et al. 2003). As well as transdifferentiation, bone marrow cells have been demonstrated to acquire non-haematopoietic phenotypes as a result of cell fusion with hepatocytes both in mice and man (Alison et al. 2000, Terada et al. 2002). Infusion of highly purified HSCs into the $FAH^{Exon\Delta5}$ model was found to rescue the phenotype as a result of cell fusion with hepatocytes (Lagasse et al. 2000). Later studies demonstrated that cell fusion occurred at a low frequency and the high level of repopulation was due to proliferation of a small number of clones (Vassilopoulos et al. 2003, Wang et al. 2003).

Bone marrow cells have also been implicated in fibrosis associated with cirrhosis (Forbes et al. 2004). In experimental models of fibrosis, bone marrow cells have been found to migrate to the injured liver, although the role that these cells play is still controversial (Houlihan and Newsome 2008). Several groups have provided evidence that bone marrow cells may contribute to fibrosis in injured livers (Forbes et al. 2004, Russo et al. 2006, Kisseleva et al. 2006), however infusion of bone marrow cells into fibrotic models has been found to reduce the extent of fibrosis and improve survival (Cho et al. 2010, Fang et al. 2004, Sakaida et al. 2004, Oyagi et al. 2006). This effect may be as a result of the expression of metalloproteases which have been implicated in the remodelling of extracellular matrix (Sakaida et al. 2004). Of note, bone marrow-derived stem cells have been demonstrated to promote hepatocyte proliferation (Yannaki et al. 2005, Piscaglia et al. 2007) which could potentially increase the rate at which the liver may recover from cirrhosis.

Mesenchymal stem cells (MSCs) have been differentiated into albumin-expressing hepatocyte-like cells (Tamagawa et al. 2007, Najimi et al. 2007, Sato et al. 2005) which have been demonstrated to repopulate the livers of murine models of fibrosis whereby they can ameliorate the extent of fibrosis and improve liver functions (Piryaei et al. 2011). These studies show great promise as a potential source of hepatocyte-like cells, however further experiments are required to demonstrate whether these cells can be grown and differentiated in sufficient numbers required to repopulate livers and that a mesenchymal origin does not predispose *in vitro* derived hepatocytes to increased matrix production and tissue fibrosis following liver injury.

Xenogeneic hepatocytes

Another possible source of primary hepatocytes may be derived from animals such as pigs. The possibility of transplanting hepatocytes from one species into another was first demonstrated by Rhim and colleagues (1995) when they transplanted rat hepatocytes into alb-uPa mice. Transplantation of rat hepatocytes into nude mice with acute liver failure could improve survival and liver function (Vidal et al. 2008). Furthermore, in an immunocompetent model of cirrhosis Nagata et al. (2003) demonstrated that porcine hepatocytes were able to prolong survival and improve liver function. This effect was greatly enhanced by treatment with the immunosuppressive FK506.

Whilst humanized porcine hepatocytes would provide an unlimited source of cells for transplantation there are concerns about the safety of transplanting animal cells into patients. The main concern is the possibility that porcine endogenous retroviruses of the animal cells will cross over the species barrier (Fishman and Patience 2004). Although most functions of the porcine liver appear to be comparable to the human liver, it is unclear how transplanted porcine hepatocytes would perform in comparison to endogenous hepatocytes when transplanted (reviewed in Ibrahim et al. 2006).

Foetal liver cells

Hepatic progenitor cells have been isolated from both foetal (Lazaro et al. 2003, Nowak et al. 2005, Schmelzer et al. 2007) and adult livers (Schmelzer et al. 2007), and enrichment of these cells has been performed using the surface markers EpCAM or $CD117^{+}$ $CD34^{+}$ lin^{-} (Nowak et al. 2005). They are bipotent cells capable of differentiating into both cholangiocytes and hepatocytes , but unlike hepatocytes, foetal liver progenitor cells (FLPCs) proliferate *in vitro*, maintaining their bipotent phenotype after prolonged culture (Lazaro et al. 2003, Schmelzer et al. 2007) Furthermore, they have been demonstrated to engraft and repopulate murine models of liver disease (Washburn et al. 2011). Experimentally, hepatic progenitor cells have been used to repopulate the livers of animal models Nowak et al. 2005, Schmelzer et al. 2007, Sandhu et al. 2001, Wang et al. 2003, Oertel et al. 2008, Haridass et al. 2009) where they have been found to acquire a more mature hepatocyte phenotype (Nowak et al. 2005). In an elegant series of experiments performed by Oertel and colleagues, this was demonstrated to be as a result of cell competition where the authors demonstrate that not only do transplanted FLPCs have much greater proliferation than the host parenchymal cells, but that the endogenous hepatocytes bordering the clusters of cells derived from the FLCPs are more susceptible to apoptosis, thereby making room for the repopulating cells (Oertel et al. 2006). Whilst FLPCs are less good than primary hepatocytes at engrafting in the liver, over time FLPCs have been found to have much greater repopulation potential as a result of increased proliferation (Sandhu et al. 2001). Interestingly, Harridass and colleagues found that FLPCs were less good at repopulating the uPA $Rag2^{-/-}$ $\gamma c^{-/-}$ mouse (Haridass et al. 2009), however this may be due to the proliferative stimulus provided by the uPA liver combined with increased engraftment.

Pluripotent stem cells and cellular transplantation of hepatocyte-like cells

The first embryonic stem cell lines were derived from the blastocysts of mice (Martin 1981, Evans and Kaufman 1981). In 1998, Thomson and colleagues reported the establishment of human ES cell lines from the inner cell mass of early stage IVF embryos (Thomson et al. 1998). These lines were found to proliferate indefinitely and contributed to the three germ layers, determined by teratoma formation in nude mice (Martin 1981, Evans and Kaufman 1981, Thomson et al. 1998). Whilst this property is desirable for regenerative medicine, it is also a double-edged sword because any undifferentiated cells could potentially lead to tumours if transplanted into patients.

The first directed differentiation of human ES cells into functional hepatocyte-like cells was performed by Rambhatla et al. (2003). Subsequently several other groups have published more efficient and robust protocols for the directed differentiation of human ES cells into functional hepatocyte-like cells (Basma et al. 2009, Agarwal et al. 2008, Cai et al. 2007, Duan et al. 2007, Hay et al. 2008a,b, Roelandt et al. 2010, Touboul et al. 2010, Zhao et al. 2009).

Groundbreaking work by Takahashi and Yamanaka demonstrated the plasticity of somatic cells when they succeeded in reprogramming both murine and human fibroblasts to an ES-like state using just four transcription factors (Takahashi et al. 2007, Takahashi and Yamanaka 2006). This generation of inducible pluripotent stem cells (iPSCs) holds great promise not only for therapeutic purposes, but also for the study of human development and genetic diseases. As such, iPSCs have successfully been isolated from patients with an array of genetic defects affecting liver function including $\alpha 1$ anti-trypsin disease, familial tyrosinaemia type I, glycogen storage disease and Crigler–Najjar Syndrome (Ghodsizadeh et al., Rashid et al. 2010). Several groups have reported the differentiation of iPSCs into functional hepatocytes using protocols similar to those used for the directed differentiation of human ES cells (Ghodsizadeh et al., Rashid et al. 2010, Sullivan et al. 2009, Si-Tayeb et al. 2010). One of the challenges facing researchers at present is the successful differentiation and engraftment of human iPSCs into animal models, without tumour formation (Basma et al. 2009, Payne et al. 2011).

In this section, we discuss the great promise that hepatic endoderm derived from human ES cells or iPSCs may hold as a means for cellular therapy as well as some of the risks associated with their use.

Safety of pluripotent stem cells and their hepatocyte-like derivatives

As their name suggest, pluripotent stem cells can differentiate into the three germ layers and therefore have the potential to become any cell in the body, as demonstrated by the ability of iPSCs to generate viable mice (Boland et al. 2009). Whilst this property is highly desirable, and forms the basis of regenerative medicine it is a double-edged sword because any undifferentiated cells form teratomas (Hanley et al. 2010), especially when transplanted into syngeneic or immunocompromised recipients, such as those on immunosuppressive drugs.

In studies transplanting ES-derived hepatocyte-like cells, several groups have reported the formation of tumours (Basma et al. 2009, Payne et al. 2011). In one study tumours were observed in both the livers and spleens, the site of delivery of the cells, that contained structures of mesodermal and endodermal, but not ectodermal origin (Payne et al. 2011) suggesting that whilst the cells had lost their pluripotent potential as a result of the initial differentiation towards the definitive endoderm, some cells either acquired a mesendodermal cell fate which was maintained through the rest of the differentiation procedure or once transplanted the cells dedifferentiated and subsequently formed tumours composed of mesodermal and endodermal cell types.

Two studies have demonstrated methods to eliminate contaminating undifferentiated cells from their cultures. Basma and colleagues (Basma et al. 2009) used the hepatocyte-specific asiaglycoprotein receptor1 to sort their cells, thereby eliminating undifferentiated cells. Whilst this strategy was successful in abrogating the formation of teratoma, the authors did report the formation of adenocarcinoma in the peritoneum of recipient mice (Basma et al. 2009). In another study, this time using murine ES cells, Li and colleagues generated hepatic progenitor-like (Li et al. 2010). Furthermore, by selecting the c-Kit$^-$ EpCAM$^+$ cells using FACS the authors demonstrated that these cells could repopulate the livers of FAH$^{Exon\Delta 5}$ mice without the generation of teratoma (Li et al. 2010)

When iPSCs were first produced, it was by reprogramming cells with retroviruses encoding the four transcription factors c-myc, KLF-4, Oct-4 and Sox-2 (MKOS). Although these viruses were shown to be silenced soon after reprogramming had occurred, when the endogenous counterparts were expressed, there is the possibility that they may become re-activated and cause tumours *in vivo*. This is particularly true of c-myc and KLF-4 which have an oncogenic capacity. Furthermore, the integration of viruses into the genome can themselves cause oncogenesis. Since these early studies, researchers have succeeded in generating iPSCs without c-Myc, however this is with greatly reduced efficiency (Huangfu et al. 2008). Furthermore, it has been demonstrated that genomic integration is not necessary for reprogramming and iPSCs can be generated using plasmids (Si-Tayeb et al. 2010, Okita et al. 2010) as well as recombinant proteins (Zhou et al. 2009) and modified RNA molecules encoding MKOS (Warren et al. 2010). Moreover, the efficiency of reprogramming has also been improved using small molecules (Huangfu et al. 2008).

Pluripotent stem cell bank

One of the risks inherent to the transplantation of cells or tissues derived from a donor is the risk of rejection by the recipient's immune system. To date, ES cell-derived progenitors are being trialled clinically to treat patients with spinal cord injury and Stargardt's muscular dystrophy (an inherited form of blindness due to photoreceptor death), both of which are immune privileged sites. For ES cell-derived progeny to have a more universal application, it will be necessary to use ES cells which are human leukocyte antigen (HLA) matched to the patient. It has been predicted by Taylor and colleagues that a bank of 150 unmatched ES cells would be sufficient to provide a partial HLA-match for up to 80% of the population. This number would reduce to a bank size of 100 if the cells are also blood group O. Furthermore, any ES cells that are homozygous for one of the three main HLA groups (HLA-A, B or DR) would be expected to provide a partial match for a greater number of patients than their heterozygous counterpart (Taylor et al. 2005).

Whilst it would be difficult to derive ES cell lines from human blastocysts for one of the HLA groups and also present blood group O antigens, an alternative may be human parthogenetic stem (PS) cell lines (Hao et al. 2009, Mai et al. 2007, Lin et al. 2007, Hwang et al. 2004, Kim et al. 2007, Revazova et al. 2007). These are generated from oocytes artificial activation (Mai et al. 2007, Lin et al. 2007, Revazova et al. 2007). These PS cell lines express markers associated with pluripotent stem cells and have a normal karyotype which has been reported to be maintained for more than a 100 passages (Mai et al. 2007). Furthermore, they have been demonstrated to generate cells of the three germ layers, both *in vitro* and *in vivo*, including functional hepatocyte-like cells (Turovets et al. 2011). Typically, PS cells are generated from mature (MII) oocytes (Mai et al. 2007, Revazova et al. 2007, Kim et al. 2007) however, as a result of meiotic recombination, almost one third of PS cells generated from MII oocytes bear heterozygous MHC molecultes (Kim et al. 2007). To overcome this, Lin and colleagues have succeeded in generating a human PS cell line that displays a high level of homozygosity throughout its genome from a one-pronuclear oocyte (Lin et al. 2007).

An alternative way to generate pluripotent stem cell repository from a more readily available source of starting material may be through iPSCs. It may be possible to screen blood donated to blood banks for samples from donors whose blood type is O and that are homozygous for one of the three main HLA groups. Since several groups have described the generation of iPSCs from blood cells (Eminli et al. 2009, Chou et al. 2011, Hanna et al. 2008) then it would be possible to generate iPSCs from those blood samples which match these criteria. Using this approach, it is conceivable that a considerable pluripotent stem cell bank could be generated using readily available source material.

Personalized medicine

Whilst the generation of a pluripotent stem cell bank could enable the production of cells—including hepatocytes- which closely match the MHC of the recipient, iPSCs have the potential to allow for cells derived from the patient to be differentiated and used to treat them. For example, it may be possible to generate iPSCs from a patient with an inborn error of metabolism and using homologous recombination correct the genetic defect before differentiating the cells towards the hepatic lineage before transplanting them back into the patient. Several groups have reported the generation of iPSCs from patients with neurodegenerative (Cheung et al. 2011, Park et al. 2008), haematopoietic (Ye et al. 2009, Raya et al. 2009) and pulmonary (Yazawa et al. 2011, Itzhaki et al. 2011, Carvajal-Vergara et al. 2010) disorders. iPSCs have also been generated from patients with inborn metabolic disorders including Familial type I tyrosinaemia, progressive familial cholestasis and Crigler-Najjar syndrome (Ghodsizadeh et al. 2010, Rashid et al. 2010). Furthermore, the generation of hepatocyte-like cells from iPSCs derived from patients with α1-anti-trypsin deficiency, glycogen storage disorder and familial hypercholesterolemia have been demonstrated to recapitulate the hepatocyte defects associated with the disorder (Rashid et al. 2010).

As a proof-of-principle to demonstrate the potential of iPSCs to treat genetic disorders, Hanna et al demonstrated that iPSCs could be used to treat a humanized mouse model of sickle cell anaemia ($h\beta^s$) by differentiating the cells towards a haematopoietic lineage before transplanting into lethally irradiated $h\beta^s$ mice (Hanna et al. 2007). This strategy was found to be successful in improving the phenotype. The authors then derived iPSCs from the $h\beta^s$ mice, corrected the genetic defect using homologous recombination and differentiating the corrected iPSCs towards haematopoietic lineage before transplanting them. This strategy was successful in improving the phenotype of the $h\beta^s$ mice and highlights the great potential of iPSCs in personalized medicine.

As has been discussed, one of the great attributes of iPSCs is that they can be generated from a patient's own of generating patient-specific iPSCs for cellular therapy thereby reducing the likelihood of rejection (Hanley et al., Chun et al. 2010). However, in a recent publication Zhao and co-workers have demonstrated that unlike ES cells, the teratoma caused by injecting iPSCs into syngeneic hosts were immunogenic (Zhao et al. 2011). The authors went on to demonstrate that this was due to the overexpression of several genes by the iPSC-derived teratoma that were not expressed by their ES-derived counterparts and that ES-derived teratoma had increased immunogenicity when these genes were expressed in ES cells. These findings could either mean that iPSC-derived cells *per se* are immunogenic when engrafted into syngenic hosts, meaning that immunosuppression would be required to abrogate rejection, or possibly that the risk of contaminating undifferentiated iPSCs forming tumours is lessened as a result of their inherent immunogenicity. Further work is needed to investigate whether iPSC-derived progeny are as immunogenic as their parent cells, or if this is reduced in the differentiated cells.

Direct reprogramming

Whilst the pluripotency of embryonic stem cells and iPSCs hold great promise for regenerative medicine, it is also a double-edged sword because of the risk of undifferentiated or partially differentiated cells forming tumours such as teratomas. As has been discussed earlier in this chapter, strategies have been described to reduce this caveat of pluripotent stem cell lines.

Reprogramming somatic cells towards iPSCs by Takahashi and Yamanaka in 2006 highlighted the plasticity of these cells when transfected with only a handful of genes (Takahashi and Yamanaka 2006). In the past few years, groups have taken the reprogramming of somatic cells to a new level by differentiating fibroblasts to neuronal cells and cardiomyocytes by transfecting cells with only a handful of genes (Ieda et al. 2010, Vierbuchen et al. 2010). Furthermore, the environment in which reprogrammed cells are cultured can affect the ultimate fate of the cells. This was highlighted by three studies which have demonstrated that fibroblasts induced to express the Yamanaka transcription factors and cultured in specific conditions can direct the reprogramming of these cells towards a cardiomyocyte (Efe et al. 2011), neural progenitor (Kim et al. 2011) or epiblast stem cell (Han et al.) phenotype, respectively. The role of the cellular microenvironment during reprogramming was elegantly highlighted by Szabo and colleagues who succeeded in generating hematopoietic progenitor-like cells through the ectopic of expression of just *oct4* in human fibroblasts and culturing in media contatining FGF2, SCF and FLT-3 (Szabo et al. 2010).

Of interest are two recent publications which demonstrated that hepatocyte-like cells could be generated directly from murine fibroblasts using a selection of hepatic-specific transcription factors (Huang et al. 2011, Sekiya and Suzuki 2011). In the manuscript by Huang and colleagues, the authors demonstrated that they could generate induced hepatocyte-like cells (iHeps) with as few as three of the transcription factors that had previously been demonstrated to be important for liver development (Huang et al. 2011). However, transduction of mouse tail tip fibroblasts with *gata3*, *hnf1a* and *foxa3* alone was insufficient to generate clones due to limited proliferation. This was overcome by inactivation of $p19^{ARF}$ which promotes hepatocyte proliferation *in vitro*. Sekiya and Suzuki also demonstrated the generation of functional iHeps by transduction of both mouse embryonic fibroblasts and adult fibroblasts using only two transcription factors, *hnf4a* alongside wither *foxa1*, *foxa2* or *foxa3*. Unlike the iHeps generated by Huang et al, the iHeps of Sekiya and Suzuki proliferated in culture without the need for further genetic manipulation. As well as demonstrating the iHeps had acquired the expression of hepatocyte-specific genes and functions, the authors of both manuscripts also demonstrated that these cells could repopulate the livers of $Fah^{Exon\Delta5}$ mice, rescuing their lethal phenotype whilst MEFs alone could not (Kim et al. 2011, Huang et al. 2011, Sekiya and Suzuki 2011). Furthermore, to further assess the therapeutic potential of iHeps Sekiya and Suzuki generated iHeps from MEFs derived from $FAH^{Exon\Delta5}$ mice by transducing them with retroviruses expressing *hnf4α*, *foxa3* and *fah*. These iheps were reported to be engraft and function in the livers of $FAH^{Exon\Delta5}$ mice, rescuing their lethal phenotype (Sekiya and Suzuki 2011).

Concluding remarks

Currently, the promise of experimental treatment of liver disorders using cellular therapy has yet to be realized clinically. However, when this does become a realistic alternative to OLT or therapy for metabolic disorders then alternative sources of cells will be required due to the shortage of donor livers. Whilst pluripotent stem cell-derived hepatocytes are a possible alternative, currently their potential is greatly hampered by lower functionality when compared to primary adult hepatocytes, safety issues as well as the regulatory requirements that would need to be met before these cells could be used clinically.

The differentiation of pluripotent stem cells and direct reprogramming of somatic cells to those with a hepatocyte phenotype is rapidly progressing so that in the near future it may be that such cells will have a functional repertoire similar to adult human hepatocytes. At this stage there will remain the challenges of achieving high levels of engraftment, whilst at the same time ensuring safety of these cells in an *in vivo* setting. Furthermore, it remains to be seen whether cells generated from iPSCs and possibly as a result of direct reprogramming are able to persist in syngenic recipient without the need for immunosuppression.

References

Agarwal S, KL Holton and R Lanza. 2008. Efficient differentiation of functional hepatocytes from human embryonic stem cells. Stem Cells. 26(5): 1117–1127.

Alison MR, R Poulsom, R Jeffery et al. 2000. Hepatocytes from non-hepatic adult stem cells. Nature. 406(6793): 257.

Allain JE, I Dagher, D Mahieu-Caputo et al. 2002. Immortalization of a primate bipotent epithelial liver stem cell. Proceedings of the National Academy of Sciences of the United States of America. 99(6): 3639–3644.

Azuma H, N Paulk, A Ranade et al. 2007. Robust expansion of human hepatocytes in Fah-/-/Rag2-/-/Il2rg-/- mice. Nat Biotechnol. 25(8): 903–910.

Basma H, A Soto-Gutierrez, GR Yannam et al. 2009. Differentiation and transplantation of human embryonic stem cell-derived hepatocytes. Gastroenterology. 136(3): 990–999.

Bissig KD, TT Le, NB Woods et al. 2007. Repopulation of adult and neonatal mice with human hepatocytes: a chimeric animal model. Proceedings of the National Academy of Sciences of the United States of America. 104(51): 20507–20511.

Boland MJ, JL Hazen, KL Nazor et al. 2009. Adult mice generated from induced pluripotent stem cells. Nature. 461(7260): 91–94.

Brazelton TR, FM Rossi, GI Keshet et al. 2000. From marrow to brain: expression of neuronal phenotypes in adult mice. Science. 290(5497): 1775–1779.

Cai J, Y Zhao, Y Liu et al. 2007. Directed differentiation of human embryonic stem cells into functional hepatic cells. Hepatology. 45(5): 1229–1239.

Carvajal-Vergara X, A Sevilla, SL D'Souza et al. 2010. Patient-specific induced pluripotent stem-cell-derived models of LEOPARD syndrome. Nature. 465(7299): 808–812.

Cheung AY, LM Horvath, D Grafodatskaya et al. 2011. Isolation of MECP2-null Rett Syndrome patient hiPS cells and isogenic controls through X-chromosome inactivation. Hum Mol Genet. 20(11): 2103–2115.

Cho KA, GW Lim, SY Joo et al. 2010. Transplantation of bone marrow cells reduces CCl(4) -induced liver fibrosis in mice. Liver Int. 31(7): 932–9.

Chou BK, P Mali, X Huang et al. 2011. Efficient human iPS cell derivation by a non-integrating plasmid from blood cells with unique epigenetic and gene expression signatures. Cell Res. 21(3): 518–529.

Chun YS, P Chaudhari and YY Jang. 2010. Applications of patient-specific induced pluripotent stem cells; focused on disease modeling, drug screening and therapeutic potentials for liver disease. Int J Biol Sci. 6(7): 796–805.

Dagher I, TH Nguyen, MT Groyer-Picard et al. 2009. Efficient hepatocyte engraftment and long-term transgene expression after reversible portal embolization in nonhuman primates. Hepatology. 49(3): 950–959.

Dawson LA and RK Ten Haken. 2005. Partial volume tolerance of the liver to radiation. Semin Radiat Oncol. 15(4): 279–283.

Delgado JP, A Parouchev, JE Allain et al. 2005. Long-term controlled immortalization of a primate hepatic progenitor cell line after Simian virus 40 T-Antigen gene transfer. Oncogene. 24(4): 541–551.

Dhawan A, RR Mitry and RD Hughes. 2006. Hepatocyte transplantation for liver-based metabolic disorders. J Inherit Metab Dis. 29(2-3): 431–435.

Duan Y, A Catana, Y Meng et al. 2007. Differentiation and enrichment of hepatocyte-like cells from human embryonic stem cells *in vitro* and *in vivo*. Stem Cells. 25(12): 3058–3068.

Efe JA, S Hilcove, J Kim et al. 2011. Conversion of mouse fibroblasts into cardiomyocytes using a direct reprogramming strategy. Nat Cell Biol. 13(3): 215–222.

Eminli S, A Foudi, M Stadtfeld et al. 2009. Differentiation stage determines potential of hematopoietic cells for reprogramming into induced pluripotent stem cells. Nature genetics. 41(9): 968–976.

Enns GM and MT Millan. 2008. Cell-based therapies for metabolic liver disease. Mol Genet Metab. 95(1-2): 3–10.

Erker L, H Azuma, AY Lee et al. 2010. Therapeutic liver reconstitution with murine cells isolated long after death. Gastroenterology. 139(3): 1019–1029.

Evans MJ and MH Kaufman. 1981. Establishment in culture of pluripotential cells from mouse embryos. Nature. 292(5819): 154–156.

Fajardo LF and TV Colby. 1980. Pathogenesis of veno-occlusive liver disease after radiation. Arch Pathol Lab Med. 104(11): 584–588.

Fang B, M Shi, L Liao et al. 2004. Systemic infusion of FLK1(+) mesenchymal stem cells ameliorate carbon tetrachloride-induced liver fibrosis in mice. Transplantation. 78(1): 83–88.

Fero ML, M Rivkin, M Tasch et al. 1996. A syndrome of multiorgan hyperplasia with features of gigantism, tumorigenesis, and female sterility in p27(Kip1)-deficient mice. Cell. 85(5): 733–744.

Ferrari G, G Cusella-De Angelis, M Coletta et al. 1998. Muscle regeneration by bone marrow-derived myogenic progenitors. Science. 279(5356): 1528–1530.

Fisher RA and SC Strom. 2006. Human hepatocyte transplantation: worldwide results. Transplantation. 82(4): 441–449.

Fishman JA and C Patience. 2004. Xenotransplantation: infectious risk revisited. Am J Transplant. 4(9): 1383–1390.

Forbes SJ, FP Russo, V Rey et al. 2004. A significant proportion of myofibroblasts are of bone marrow origin in human liver fibrosis. Gastroenterology. 126(4): 955–963.

Fox IJ, JR Chowdhury, SS Kaufman et al. 1998. Treatment of the Crigler-Najjar syndrome type I with hepatocyte transplantation. N Engl J Med. 338(20): 1422–1426.

Fu PP, Q Xia, G Lin et al. 2004. Pyrrolizidine alkaloids—genotoxicity, metabolism enzymes, metabolic activation, and mechanisms. Drug Metab Rev. 36(1): 1–55.

Ghodsizadeh A, A Taei, M Totonchi et al. 2011. Generation of liver disease-specific induced pluripotent stem cells along with efficient differentiation to functional hepatocyte-like cells. Stem Cell Rev. 6(4): 622–632.

Grompe M, M Al-Dhalimy, M Finegold et al. 1993. Loss of fumarylacetoacetate hydrolase is responsible for the neonatal hepatic dysfunction phenotype of lethal albino mice. Genes Dev. 7(12A): 2298–2307.

Grompe M, S Lindstedt, M Al-Dhalimy et al. 1995. Pharmacological correction of neonatal lethal hepatic dysfunction in a murine model of hereditary tyrosinaemia type I. Nature genetics. 10(4): 453–460.

Grompe M. 2006. Principles of therapeutic liver repopulation. J Inherit Metab Dis. 29(2-3): 421–425.

Grossman M, DJ Rader, DW Muller et al. 1995. A pilot study of ex vivo gene therapy for homozygous familial hypercholesterolaemia. Nature medicine. 1(11): 1148–1154.

Grossman M, SE Raper, K Kozarsky et al. 1994. Successful ex vivo gene therapy directed to liver in a patient with familial hypercholesterolaemia. Nature genetics. 6(4): 335–341.

Groth CG, B Arborgh, C Bjorken et al. 1977. Correction of hyperbilirubinemia in the glucuronyltransferase-deficient rat by intraportal hepatocyte transplantation. Transplant Proc. 9(1): 313–316.

Guha C, A Sharma, S Gupta et al. 1999. Amelioration of radiation-induced liver damage in partially hepatectomized rats by hepatocyte transplantation. Cancer research. 59(23): 5871–5874.

Guha C, NJ Deb, BS Sappal et al. 2001. Amplification of engrafted hepatocytes by preparative manipulation of the host liver. Artif Organs. 25(7): 522–528.

Gupta S, P Rajvanshi, R Sokhi et al. 1999. Entry and integration of transplanted hepatocytes in rat liver plates occur by disruption of hepatic sinusoidal endothelium. Hepatology. 29(2): 509–519.

Han DW, B Greber, G Wu et al. 2011. Direct reprogramming of fibroblasts into epiblast stem cells. Nat Cell Biol. 13(1): 66–71.

Hanley J, G Rastegarlari and AC Nathwani. 2010. An introduction to induced pluripotent stem cells. British journal of haematology. 151(1): 16–24.

Hanna J, M Wernig, S Markoulaki et al. 2007. Treatment of sickle cell anemia mouse model with iPS cells generated from autologous skin. Science. 318(5858): 1920–1923.

Hanna J, S Markoulaki, P Schorderet et al. 2008. Direct reprogramming of terminally differentiated mature B lymphocytes to pluripotency. Cell. 133(2): 250–264.

Hao J, W Zhu, C Sheng et al. 2009. Human parthenogenetic embryonic stem cells: one potential resource for cell therapy. Sci China C Life Sci. 52(7): 599–602.

Haridass D, Q Yuan, PD Becker et al. 2009. Repopulation Efficiencies of Adult Hepatocytes, Fetal Liver Progenitor Cells, and Embryonic Stem Cell-Derived Hepatic Cells in Albumin-Promoter-Enhancer Urokinase-Type Plasminogen Activator Mice. Am J Pathol. 175(4): 1483–92.

Hay DC, D Zhao, J Fletcher et al. 2008a. Efficient differentiation of hepatocytes from human embryonic stem cells exhibiting markers recapitulating liver development *in vivo*. Stem Cells. 26(4): 894–902.

Hay DC, J Fletcher, C Payne et al. 2008b. Highly efficient differentiation of hESCs to functional hepatic endoderm requires ActivinA and Wnt3a signaling. Proceedings of the National Academy of Sciences of the United States of America. 105(34): 12301–12306.

Houlihan DD and PN Newsome. 2008. Critical review of clinical trials of bone marrow stem cells in liver disease. Gastroenterology. 135(2): 438–450.

Huang P, Z He, S Ji et al. 2011. Induction of functional hepatocyte-like cells from mouse fibroblasts by defined factors. Nature. 475(7356): 386–9.

Huangfu D, K Osafune, R Maehr et al. 2008. Induction of pluripotent stem cells from primary human fibroblasts with only Oct4 and Sox2. Nat Biotechnol. 26(11): 1269–1275.

Hwang WS, YJ Ryu and JH Park et al. 2004. Evidence of a pluripotent human embryonic stem cell line derived from a cloned blastocyst. Science. 303(5664): 1669–1674.

Ibrahim Z, J Busch, M Awwad et al. 2006. Selected physiologic compatibilities and incompatibilities between human and porcine organ systems. Xenotransplantation. 13(6): 488–499.

Ieda M, JD Fu, P Delgado-Olguin et al. 2010. Direct reprogramming of fibroblasts into functional cardiomyocytes by defined factors. Cell. 142(3): 375–386.

Ishikawa F, CJ Drake, S Yang et al. 2003. Transplanted human cord blood cells give rise to hepatocytes in engrafted mice. Ann N Y Acad Sci. 996: 174–185.

Itzhaki I, L Maizels, I Huber et al. 2011. Modelling the long QT syndrome with induced pluripotent stem cells. Nature. 471(7337): 225–229.

Joseph B, H Malhi, KK Bhargava et al. 2002. Kupffer cells participate in early clearance of syngeneic hepatocytes transplanted in the rat liver. Gastroenterology. 123(5): 1677–1685.

Joseph B, V Kumaran, E Berishvili et al. 2006. Monocrotaline promotes transplanted cell engraftment and advances liver repopulation in rats via liver conditioning. Hepatology. 44(6): 1411–1420.

Karnezis AN, M Dorokhov, M Grompe et al. 2001. Loss of p27(Kip1) enhances the transplantation efficiency of hepatocytes transferred into diseased livers. J Clin Invest. 108(3): 383–390.

Kawahara T, C Toso, DN Douglas et al. 2010a. Factors affecting hepatocyte isolation, engraftment, and replication in an *in vivo* model. Liver Transpl. 16(8): 974–982.

Kawahara T, DN Douglas, J Lewis et al. 2010b. Critical role of natural killer cells in the rejection of human hepatocytes after xenotransplantation into immunodeficient mice. Transpl Int.

Kawahara T, S Kasai, H Yagita et al. 1997. Involvement of Fas and Fas ligand interaction in allogeneic hepatocyte rejection in the spleen. Transplant Proc. 29(4): 2187–2188.

Kim BH, SR Sung, EH Choi et al. 2000. Dedifferentiation of conditionally immortalized hepatocytes with long-term *in vitro* passage. Exp Mol Med. 32(1): 29–37.

Kim J, JA Efe, S Zhu et al. 2011. Direct reprogramming of mouse fibroblasts to neural progenitors. Proceedings of the National Academy of Sciences of the United States of America. 108(19): 7838–7843.

Kim K, K Ng, PJ Rugg-Gunn et al. 2007. Recombination signatures distinguish embryonic stem cells derived by parthenogenesis and somatic cell nuclear transfer. Cell Stem Cell. 1(3): 346–352.

Kim K, P Lerou, A Yabuuchi et al. 2007. Histocompatible embryonic stem cells by parthenogenesis. Science. 315(5811): 482–486.

Kisseleva T, H Uchinami, N Feirt et al. 2006. Bone marrow-derived fibrocytes participate in pathogenesis of liver fibrosis. J Hepatol. 45(3): 429–438.

Kiyokawa H, RD Kineman, KO Manova-Todorova et al. 1996. Enhanced growth of mice lacking the cyclin-dependent kinase inhibitor function of p27(Kip1). Cell. 85(5): 721–732.

Kobayashi N, T Fujiwara, KA Westerman et al. 2000. Prevention of acute liver failure in rats with reversibly immortalized human hepatocytes. Science. 287(5456): 1258–1262.

Kocken JM, IH Borel Rinkes, AM Bijma et al. 1996. Correction of an inborn error of metabolism by intraportal hepatocyte transplantation in a dog model. Transplantation. 62(3): 358–364.

Krohn N, S Kapoor, Y Enami et al. 2009. Hepatocyte transplantation-induced liver inflammation is driven by cytokines-chemokines associated with neutrophils and Kupffer cells. Gastroenterology. 136(5): 1806–1817.

Kumaran V, B Joseph, D Benten et al. 2005. Integrin and extracellular matrix interactions regulate engraftment of transplanted hepatocytes in the rat liver. Gastroenterology. 129(5): 1643–1653.

Laconi E, R Oren, DK Mukhopadhyay et al. 1998. Long-term, near-total liver replacement by transplantation of isolated hepatocytes in rats treated with retrorsine. Am J Pathol. 153(1): 319–329.

Lagasse E, H Connors, M Al-Dhalimy et al. 2000. Purified hematopoietic stem cells can differentiate into hepatocytes *in vivo*. Nature medicine. 6(11): 1229–1234.

Lazaro CA, EJ Croager, C Mitchell et al. 2003. Establishment, characterization, and long-term maintenance of cultures of human fetal hepatocytes. Hepatology. 38(5): 1095–1106.

Lee JS, D Semela, J Iredale et al. 2007. Sinusoidal remodeling and angiogenesis: a new function for the liver-specific pericyte? Hepatology. 45(3): 817–825.

Li F, P Liu, C Liu et al. 2010. Hepatoblast-like progenitor cells derived from embryonic stem cells can repopulate livers of mice. Gastroenterology. 139(6): 2158–2169 e2158.

Lin G, Q OuYang, X Zhou et al. 2007. A highly homozygous and parthenogenetic human embryonic stem cell line derived from a one-pronuclear oocyte following *in vitro* fertilization procedure. Cell Res. 17(12): 999–1007.

Mai Q, Y Yu, T Li et al. 2007. Derivation of human embryonic stem cell lines from parthenogenetic blastocysts. Cell Res. 17(12): 1008–1019.

Makowka L, LE Rotstein, RE Falk et al. 1980. Allogeneic and xenogeneic hepatocyte transplantation in experimental hepatic failure. Transplantation. 30(6): 429–435.

Malhi H, P Annamaneni, S Slehria et al. 2002. Cyclophosphamide disrupts hepatic sinusoidal endothelium and improves transplanted cell engraftment in rat liver. Hepatology. 36(1): 112–121.

Martin GR. 1981. Isolation of a pluripotent cell line from early mouse embryos cultured in medium conditioned by teratocarcinoma stem cells. Proceedings of the National Academy of Sciences of the United States of America. 78(12): 7634–7638.

Matas AJ, DE Sutherland, MW Steffes et al. 1976. Hepatocellular transplantation for metabolic deficiencies: decrease of plasms bilirubin in Gunn rats. Science. 192(4242): 892–894.

Mercer DF, DE Schiller, JF Elliott et al. 2001. Hepatitis C virus replication in mice with chimeric human livers. Nature medicine. 7(8): 927–933.

Mezey E, KJ Chandross, G Harta et al. 2000. Turning blood into brain: cells bearing neuronal antigens generated *in vivo* from bone marrow. Science. 290(5497): 1779–1782.

Mignon A, JE Guidotti, C Mitchell et al. 1998. Selective repopulation of normal mouse liver by Fas/CD95-resistant hepatocytes. Nature medicine. 4(10): 1185–1188.

Mitchell C, A Mignon, JE Guidotti et al. 2000. Therapeutic liver repopulation in a mouse model of hypercholesterolemia. Hum Mol Genet. 9(11): 1597–1602.

Mitchell C, VO Mallet, JE Guidotti et al. 2002. Liver repopulation by Bcl-x(L) transgenic hepatocytes. Am J Pathol. 160(1): 31–35.

Nagata H, M Ito, J Cai et al. 2003. Treatment of cirrhosis and liver failure in rats by hepatocyte xenotransplantation. Gastroenterology. 124(2): 422–431.

Najimi M, DN Khuu, PA Lysy et al. 2007. Adult-derived human liver mesenchymal-like cells as a potential progenitor reservoir of hepatocytes? Cell Transplant. 16(7): 717–728.

Nakamura J, T Okamoto, IK Schumacher et al. 1997. Treatment of surgically induced acute liver failure by transplantation of conditionally immortalized hepatocytes. Transplantation. 63(11): 1541–1547.

Nakayama K, N Ishida, M Shirane et al. 1996. Mice lacking p27(Kip1) display increased body size, multiple organ hyperplasia, retinal dysplasia, and pituitary tumors. Cell. 85(5): 707–720.

Newsome PN, I Johannessen, S Boyle et al. 2003. Human cord blood-derived cells can differentiate into hepatocytes in the mouse liver with no evidence of cellular fusion. Gastroenterology. 124(7): 1891–1900.

Nowak G, BG Ericzon, S Nava et al. 2005. Identification of expandable human hepatic progenitors which differentiate into mature hepatic cells *in vivo*. Gut. 54(7): 972–979.

Oertel M, A Menthena, MD Dabeva et al. 2006. Cell competition leads to a high level of normal liver reconstitution by transplanted fetal liver stem/progenitor cells. Gastroenterology. 130(2): 507–520.

Oertel M, A Menthena, YQ Chen et al. 2008. Purification of fetal liver stem/progenitor cells containing all the repopulation potential for normal adult rat liver. Gastroenterology. 134(3): 823–832.

Okita K, H Hong, K Takahashi et al. 2010. Generation of mouse-induced pluripotent stem cells with plasmid vectors. Nat Protoc. 5(3): 418–428.

Overturf K, M Al-Dhalimy, M Finegold et al. 1999. The repopulation potential of hepatocyte populations differing in size and prior mitotic expansion. Am J Pathol. 155(6): 2135–2143.

Oyagi S, M Hirose, M Kojima et al. 2006. Therapeutic effect of transplanting HGF-treated bone marrow mesenchymal cells into CCl4-injured rats. J Hepatol. 44(4): 742–748.

Pan X, W Du, X Yu et al. 2010. Establishment and characterization of immortalized porcine hepatocytes for the study of hepatocyte xenotransplantation. Transplant Proc. 42(5): 1899–1906.

Park IH, N Arora, H Huo et al. 2008. Disease-specific induced pluripotent stem cells. Cell. 134(5): 877–886.

Payne CM, K Samuel, A Pryde et al. 2011. Persistence of functional hepatocyte-like cells in immune-compromised mice. Liver Int. 31(2): 254–262.

Piryaei A, MR Valojerdi, M Shahsavani et al. 2011. Differentiation of bone marrow-derived mesenchymal stem cells into hepatocyte-like cells on nanofibers and their transplantation into a carbon tetrachloride-induced liver fibrosis model. Stem Cell Rev. 7(1): 103–118.

Piscaglia AC, TD Shupe, SH Oh et al. 2007. Granulocyte-colony stimulating factor promotes liver repair and induces oval cell migration and proliferation in rats. Gastroenterology. 133(2): 619–631.

Ponder KP, S Gupta, F Leland et al. 1991. Mouse hepatocytes migrate to liver parenchyma and function indefinitely after intrasplenic transplantation. Proceedings of the National Academy of Sciences of the United States of America. 88(4): 1217–1221.

Rambhatla L, CP Chiu, P Kundu et al. 2003. Generation of hepatocyte-like cells from human embryonic stem cells. Cell Transplant. 12(1): 1–11.
Raper SE, M Grossman, DJ Rader et al. 1996. Safety and feasibility of liver-directed ex vivo gene therapy for homozygous familial hypercholesterolemia. Ann Surg. 223(2): 116–126.
Rashid ST, S Corbineau, N Hannan et al. 2010. Modeling inherited metabolic disorders of the liver using human induced pluripotent stem cells. J Clin Invest. 120(9): 3127–3136.
Raya A, I Rodriguez-Piza, G Guenechea et al. 2009. Disease-corrected haematopoietic progenitors from Fanconi anaemia induced pluripotent stem cells. Nature. 460(7251): 53–59.
Revazova ES, NA Turovets, OD Kochetkova et al. 2007. Patient-specific stem cell lines derived from human parthenogenetic blastocysts. Cloning Stem Cells. 9(3): 432–449.
Rhim JA, EP Sandgren, JL Degen et al. 1994. Replacement of diseased mouse liver by hepatic cell transplantation. Science. 263(5150): 1149–1152.
Rhim JA, EP Sandgren, RD Palmiter et al. 1995. Complete reconstitution of mouse liver with xenogeneic hepatocytes. Proceedings of the National Academy of Sciences of the United States of America. 92(11): 4942–4946.
Roelandt P, KA Pauwelyn, P Sancho-Bru et al. 2010. Human embryonic and rat adult stem cells with primitive endoderm-like phenotype can be fated to definitive endoderm, and finally hepatocyte-like cells. PLoS One. 5(8): e12101.
Russo FP, MR Alison, BW Bigger et al. 2006. The bone marrow functionally contributes to liver fibrosis. Gastroenterology. 130(6): 1807–1821.
Sakaida I, S Terai, N Yamamoto et al. 2004. Transplantation of bone marrow cells reduces CCl4-induced liver fibrosis in mice. Hepatology. 40(6): 1304–1311.
Sandhu JS, PM Petkov, MD Dabeva et al. 2001. Stem cell properties and repopulation of the rat liver by fetal liver epithelial progenitor cells. Am J Pathol. 159(4): 1323–1334.
Sato Y, H Araki, J Kato et al. 2005. Human mesenchymal stem cells xenografted directly to rat liver are differentiated into human hepatocytes without fusion. Blood. 106(2): 756–763.
Schmelzer E, L Zhang, A Bruce et al. 2007. Human hepatic stem cells from fetal and postnatal donors. The Journal of experimental medicine. 204(8): 1973–1987.
Schumacher IK, T Okamoto, BH Kim et al. 1996. Transplantation of conditionally immortalized hepatocytes to treat hepatic encephalopathy. Hepatology. 24(2): 337–343.
Sekiya S and A Suzuki. 2011. Direct conversion of mouse fibroblasts to hepatocyte-like cells by defined factors. Nature. 475(7356): 390–3.
Selden C, D Calnan, N Morgan et al. 1995. Histidinemia in mice: a metabolic defect treated using a novel approach to hepatocellular transplantation. Hepatology. 21(5): 1405–1412.
Sherr CJ and JM Roberts. 1995. Inhibitors of mammalian G1 cyclin-dependent kinases. Genes Dev. 9(10): 1149–1163.
Sherr CJ and JM Roberts. 1999. CDK inhibitors: positive and negative regulators of G1-phase progression. Genes Dev. 13(12): 1501–1512.
Shneider BL. 2002. Pediatric liver transplantation in metabolic disease: clinical decision making. Pediatr Transplant. 6(1): 25–29.
Si-Tayeb K, FK Noto, A Sepac et al. 2010. Generation of human induced pluripotent stem cells by simple transient transfection of plasmid DNA encoding reprogramming factors. BMC Dev Biol. 10: 81.
Si-Tayeb K, FK Noto, M Nagaoka et al. 2010. Highly efficient generation of human hepatocyte-like cells from induced pluripotent stem cells. Hepatology. 51(1): 297–305.
Smets FN, Y Chen, LJ Wang et al. 2002. Loss of cell anchorage triggers apoptosis (anoikis) in primary mouse hepatocytes. Mol Genet Metab. 75(4): 344–352.
Sommer BG, DE Sutherland, RL Simmons et al. 1980. Hepatocellular transplantation for experimental ischemic acute liver failure in dogs. J Surg Res. 29(4): 319–325.
Song X, Y Guo, S Duo et al. 2009. A mouse model of inducible liver injury caused by tet-on regulated urokinase for studies of hepatocyte transplantation. Am J Pathol. 175(5): 1975–1983.

Sullivan GJ, DC Hay, IH Park et al. 2009. Generation of functional human hepatic endoderm from human induced pluripotent stem cells. Hepatology. 51(1): 329–35.

Szabo E, S Rampalli, RM Risueno et al. 2010. Direct conversion of human fibroblasts to multilineage blood progenitors. Nature. 468(7323): 521–526.

Tada K, N Roy-Chowdhury, V Prasad et al. 1998. Long-term amelioration of bilirubin glucuronidation defect in Gunn rats by transplanting genetically modified immortalized autologous hepatocytes. Cell Transplant. 7(6): 607–616.

Takahashi K and S Yamanaka. 2006. Induction of pluripotent stem cells from mouse embryonic and adult fibroblast cultures by defined factors. Cell. 126(4): 663–676.

Takahashi K, K Tanabe, M Ohnuki et al. 2007. Induction of pluripotent stem cells from adult human fibroblasts by defined factors. Cell. 131(5): 861–872.

Takahashi M, NJ Deb, Y Kawashita et al. 2003. A novel strategy for *in vivo* expansion of transplanted hepatocytes using preparative hepatic irradiation and FasL-induced hepatocellular apoptosis. Gene Ther. 10(4): 304–313.

Tamagawa T, S Oi, I Ishiwata et al. 2007. Differentiation of mesenchymal cells derived from human amniotic membranes into hepatocyte-like cells *in vitro*. Hum Cell. 20(3): 77–84.

Taylor CJ, EM Bolton, S Pocock et al. 2005. Banking on human embryonic stem cells: estimating the number of donor cell lines needed for HLA matching. Lancet. 366(9502): 2019–2025.

Terada N, T Hamazaki, M Oka et al. 2002. Bone marrow cells adopt the phenotype of other cells by spontaneous cell fusion. Nature. 416(6880): 542–545.

Thomson JA, J Itskovitz-Eldor, SS Shapiro et al. 1998. Embryonic stem cell lines derived from human blastocysts. Science. 282(5391): 1145–1147.

Touboul T, NR Hannan, S Corbineau et al. 2010. Generation of functional hepatocytes from human embryonic stem cells under chemically defined conditions that recapitulate liver development. Hepatology. 51(5): 1754–1765.

Turovets N, J Fair, R West et al. 2011. Derivation of High-Purity Definitive Endoderm from Human Parthenogenetic Stem Cells using an *in vitro* Analog of the Primitive Streak. Cell Transplant.

Vassilopoulos G, PR Wang and DW Russell. 2003. Transplanted bone marrow regenerates liver by cell fusion. Nature. 422(6934): 901–904.

Vidal I, N Blanchard, E Alexandre et al. 2008. Improved xenogenic hepatocyte implantation into nude mouse liver parenchyma with acute liver failure when followed by repeated anti-Fas antibody (Jo2) treatment. Cell Transplant. 17(5): 507–524.

Vierbuchen T, A Ostermeier, ZP Pang et al. 2010. Direct conversion of fibroblasts to functional neurons by defined factors. Nature. 463(7284): 1035–1041.

Wang J, W Li, J Min et al. 2004. Intrasplenic transplantation of allogeneic hepatocytes modified by BCL-2 gene protects rats from acute liver failure. Transplant Proc. 36(10): 2924–2926.

Wang X, H Willenbring, Y Akkari et al. 2003. Cell fusion is the principal source of bone-marrow-derived hepatocytes. Nature. 422(6934): 897–901.

Wang X, M Foster, M Al-Dhalimy et al. 2003. The origin and liver repopulating capacity of murine oval cells. Proceedings of the National Academy of Sciences of the United States of America. 100 Suppl 1: 11881–11888.

Warren L, PD Manos, T Ahfeldt et al. 2010. Highly efficient reprogramming to pluripotency and directed differentiation of human cells with synthetic modified mRNA. Cell Stem Cell. 7(5): 618–630.

Washburn ML, MT Bility, L Zhang et al. 2011. A humanized mouse model to study hepatitis C virus infection, immune response, and liver disease. Gastroenterology. 140(4): 1334–1344.

Wiederkehr JC, GT Kondos and R Pollak. 1990. Hepatocyte transplantation for the low-density lipoprotein receptor-deficient state. A study in the Watanabe rabbit. Transplantation. 50(3): 466–471.

Wilhelm A, I Leister, P Sabandal et al. 2004. Acute impairment of hepatic microcirculation and recruitment of nonparenchymal cells by intrasplenic hepatocyte transplantation. J Pediatr Surg. 39(8): 1214–1219.
Wu YM, B Joseph, E Berishvili et al. 2008. Hepatocyte transplantation and drug-induced perturbations in liver cell compartments. Hepatology. 47(1): 279–287.
Yamanouchi K, H Zhou, N Roy-Chowdhury et al. 2009. Hepatic irradiation augments engraftment of donor cells following hepatocyte transplantation. Hepatology. 49(1): 258–267.
Yannaki E, E Athanasiou, A Xagorari et al. 2005. G-CSF-primed hematopoietic stem cells or G-CSF per se accelerate recovery and improve survival after liver injury, predominantly by promoting endogenous repair programs. Exp Hematol. 33(1): 108–119.
Yazawa M, B Hsueh, X Jia et al. 2011. Using induced pluripotent stem cells to investigate cardiac phenotypes in Timothy syndrome. Nature. 471(7337): 230–234.
Ye Z, H Zhan, P Mali et al. 2009. Human-induced pluripotent stem cells from blood cells of healthy donors and patients with acquired blood disorders. Blood. 114(27): 5473–5480.
Zhao D, S Chen, J Cai et al. 2009. Derivation and characterization of hepatic progenitor cells from human embryonic stem cells. PLoS One. 4(7): e6468.
Zhao T, ZN Zhang, Z Rong et al. 2011. Immunogenicity of induced pluripotent stem cells. Nature. 474(7350): 212–215.
Zhou H, S Wu, JY Joo et al. 2009. Generation of induced pluripotent stem cells using recombinant proteins. Cell Stem Cell. 4(5): 381–384.

10

Advances in Cell Therapy for Liver Disease

Luke Boulter

Introduction

There is an ever increasing mismatch between the number of patients who would benefit from liver transplantation and the number of available donor organs. Furthermore a liver transplant is a highly involved surgical procedure which would require the recipient is immunosuppressed for the remainder of their lives. As such there is therefore a concerted global impetus to develop alternative treatment strategies for the treatment of severe liver disease. Along with this has emerged a growing body of data investigating novel cell types in the regeneration of the adult liver. The work in this field is still very much in its infancy; however there are numerous publications looking at both preclinical animal models as well as small, non-controlled and larger randomized, controlled human trials seeking to restore a functional hepatic mass in a plethora of human disease pathologies.

Cell and stem cell therapy for acute liver failure

Currently the only therapy for acute liver failure (ALF) is the use of orthotopic liver transplantation (OLT), however due to a lack of suitable donor livers many potential recipients die waiting for these tissues to become available. In this context it is conceivable that transplantation of

Research Fellow, Queens Medical Research Institute, 47 Little France Crescent, Edinburgh, EH16 4TJ, UK; Email: lboulter@staffmail.ed.ac.uk the metabolic component of the liver, the hepatocyte, could be useful for treatment, or sustain the patient until a suitable donor organ becomes available. The widespread use of hepatocyte transplantation is limited by the availability of primary human hepatocytes. Currently hepatocytes which could be used for transplantation come from orthotopic livers which are deemed unsuitable for transplant. As these organs have been rejected for surgical replacement due to their low quality it is not surprising that the hepatocytes which are derived from these organs are often of a low quality, and as a consequence are not always successfully used in a hepatocyte replacement strategy. As large numbers of high quality hepatocytes would be required for the successful transplantation and reconstitution of the liver, stem cell-derived hepatocytes may hold great promise as stem cell numbers can be greatly enhanced and increased through well defined and well documented *in vitro* methods. These large scale cultures can then be differentiated into hepatocytes and used for transplantation, supplying a large number of high quality, pathogen free donor cells.

During acute phase disease in ALF the micro-architecture of the liver is highly disrupted and the environment into which cadaveric or stem-cell derived hepatocytes may be transplanted will not be conducive for maintenance of hepatocyte function. As such extra-corporeal liver devices which are charged with donor hepatocytes could be utilized to mimic the liver (McKenzie et al. 2008). Extra-corporeal liver devices would be required to detoxify plasma from a liver failure patent which could be highly cytotoxic. As such it would be important in this situation that the hepatocytes contained within this device are metabolically and synthetically close to normal hepatocytes. Moreover in ALF there is rapid requirement for application of hepatocyte therapy and as such these cells must be mobilized quickly from cryopreservation, as a long term strategy would invariably be of small value to these patients. Recent advances in the derivation of hepatocytes from both Embryonic Stem Cells (ESCs) and Induced Pluripotent Stem (iPS) cells appear to have the potential for large scale expansion and would be capable of differentiation into hepatocytes for clinical use. Given our current understanding of these stem cells there would still be significant concern regarding the transplantation of these cells into recipients as contamination from ESCs or iPS cells is known to readily form teratoma and as such could lead to further clinical complications post hepatocyte transplantation.

Auxiliary liver transplantation in the management of patients with acute liver failure (ALF) gives hope that the engraftment of small amounts of liver parenchyma can allow the native liver to regenerate sufficiently for clinical recovery to occur whilst the critical liver functions are being maintained by the small amount of donor tissue. The same principal holds for hepatocyte transplantation, i.e., to provide adequate metabolic support until the endogenous liver is able to regenerate itself or even as a bridge to orthotopic liver transplantation (Fisher and Strom 2006). Whilst no randomized controlled trials exist to explore the efficacy of hepatocyte transplantation into patients some reports indicate that the hepatocyte infusion (if the cells are given in great enough quantities) are having a positive effect after they have engrafted in the recipient and can exert their metabolic function *in vivo*. Patients receiving the cells have either been bridged to transplantation or even survived long-term. Other studies into this therapeutic strategy are less encouraging, for example in a study by Bilir et al. (2000) five patients with ALF who were not candidates for OLT underwent hepatocyte transplantation. The authors describe a transient improvement in clinical condition however there was no increase in long-term survival.

Whilst interest has obviously centred upon the use of hepatocytes for the metabolic support of acute liver failure, recent work suggests that targetting other cell populations may also have therapeutic potential. An interesting study from Belardinelli et al. (2008) have used intra-portal injections of bone marrow (BM) derived mononuclear cells to reduce liver necrosis and improve survival in an acetaminophen-induced model of acute liver failure in rats. This is an exciting if preliminary report as to date cell therapy for liver failure has largely focused on providing sufficient numbers of healthy hepatocytes, it may be that modulation of the tissue response to injury or regenerative response may be equally useful in this form of liver injury. If liver regeneration can be sufficiently supported then a return to compensated liver function can be anticipated.

Cell and stem cell therapy for metabolic diseases

The ability of the liver to regenerate itself is remarkable; however for patients suffering a suite of genetic or metabolic diseases such as tyrosinaemia type 1 (deficiency of the hepatocyte enzyme fumarylacetoacetate hydrolase, the final enzyme of the catabolic pathway of tyrosine) and Crigler–Najjar syndrome type 1A (lack of functional UDP glucuronosyltransferase 1) an exogenous "non-mutant" source of hepatocytes would be desirable. These two metabolic syndromes as rodent models of these conditions have been identified by me and are extremely informative in the understanding of the requirements of successful cell therapy strategies for liver disease in a context which would promote cell uptake, engraftment and metabolic restoration where incomplete replacement of the liver may have therapeutic effect if a threshold of enzymatic function is surpassed. Moreover Overturf et al. (1996) demonstrated that in a mouse model of tyrosinaemia (the FAH deficient mouse) that hepatocyte transplant was curative as the donor hepatocytes have a survival, and therefore selection advantage over the defective host cells (that is to say they are clonogenic *in vivo*); over time the liver is predominantly comprised of the donor hepatocytes.

Other important hurdles must be overcome in order to work-up an effective cell therapy for liver disease, and some of the basic principles of these processes such as administration and cellular preservation have been demonstrated in rodent models (Slehria et al. 2002, Oertel et al. 2006).

Small pioneering clinical studies have used hepatocyte transplantation in the treatment of Crigler Najjar syndrome type 1 (Fox et al. 1998). In this study, partial phenotypic improvement was observed in the recipient. Glycogen storage disease type 1A has been treated by infusion of 2 billion hepatocytes (Muraca et al. 2002), again partial, although clinically significant, phenotype correction was seen. Although immunosuppression was used, the correction was not permanent. Sokal et al. (2003) used hepatocyte transplantation in a 4-yr-old girl with peroxisomal biogenesis disease (infantile Refsum disease), here an approximately 40% improvement in metabolic function was seen persisting at 18 mon. Furthermore, hepatic chimerism was demonstrated between the donor and endogenous cells.

Cell sources of hepatocytes

Endogenous adult hepatocytes have a multitude of functions which depends entirely on both their transcriptional context and also spatial location within the adult liver, whether these hepatocytes are close to portal or venous circulation plays a critical role in the maintenance and diversity of their synthetic function. As such it is difficult to precisely replicate these characteristics using *in vitro* differentiated cells; where the environmental context cannot be completely replicated. As these cells often fall short of functional hepatocytes, they are termed "hepatocyte-like" and this phenotype, and the markers often associated with them do not guarantee cells will emulate the highly specialized synthetic, metabolic and storage roles of hepatocytes. Phenotypic stability and appropriate proliferation over time must be firmly established before stem cell derived "hepatocyte-like" cells are used in the clinic as both of these factors could affect the ability of the treated liver to cope with the normal homeostatic demands of the patient as well as the potential for these cells to become oncogenic (Payne et al. 2011).

Embryonic stem cells

The human blastocyst develops some 4–5 d after fertilization. At this point the embryo contains approximately 50–150 cells and is separated into the trophoblast, which will become the placental tissues, and the inner cell mass which ultimately becomes the embryo-proper. Embryonic stem cells (ESCs) are derived from the dissociation of the inner cell mass and are pluripotent meaning they can give rise to the three germ layers: endoderm (including liver), mesoderm and ectoderm. The ability to divide in a limitless fashion and to differentiate along selected and controllable lineages has led to these cells being toted as the gold standard in regenerative medicine; the fact that these cells too can be efficiently differentiated into hepatocyte-like cells that possess many of the synthetic and metabolic properties of mature hepatocytes is also desirable. Differentiation of the hESC into hepatocytes has been led through an "Evo-Devo" approach, which has defined the factors conserved in the differentiation of hepatocytes during ontogeny and then apply these to the *in vitro* situation. In this context there is a clear requirement to go through a stage-wise approach where ES cells form hepatic endoderm and then are matured into hepatocytes in a similar way to the establishment of the liver bud and its maturation in the developing mammalian embryo (Hay et al. 2008, Hay et al. 2011, Medine et al. 2011). The long debated ethical issues surrounding the use of ES cells for research and therapy are still hotly contested by both sides; however the biological constraints upon the use of ES cells for therapy is also significant. ES cells have the ability to form teratomas *in vivo* as such transplantation of these cells or their progeny would entail risk to the recipient moreover ES-derived hepatocytes are non-autologous and therefore require immunosuppression akin to that of an adult hepatocyte transplant.

Induced pluripotent stem cells

The recent development of induced pluripotent cells technology has transformed the usability of ES technology and tools, and potentially offers an ethically sound alternative to the use of ES cells in both basic research and therapeutic treatment. Yamanaka and colleagues demonstrated that using donor murine fibroblasts transfected with viral Oct4, Klf4, Sox2 and c-Myc these cells underwent reprogramming and reverted from their differentiated phenotype into a naïve, ES-like fate (Takahashi and Yamanaka 2006). iPS cells could in turn differentiate again into any of the three germ layers and also demonstrated this tri-lineage potential in mouse. Importantly this has been shown, using the same techniques, in human cells (Takahashi et al. 2007). Concerns over the oncogenic potential of c-Myc, led to it being excluded from further work, where reprogramming remained intact—albeit at a lower frequency than originally reported (Nakagawa et al. 2008). There are of course safety issues regarding the use of integrating viral vectors to deliver the reprogramming genes. Recent studies have indicated that reprogramming can be achieved using plasmids and other systems to transiently express the genes and induce pluripotency without genomic integration (Kaji et al. 2009, Okita et al. 2008). iPS however does not mark final stage in derivation of hepatocytes for therapy however. Despite these cells recently being demonstrated as capable of forming hepatocytes *in vitro* (Sullivan et al. 2010, Si-Tayeb et al. 2010) they also form teratomas upon transplant. Moreover the immunological benefits of reprogrammed donor cells have recently been investigated and findings suggest that iPS cells, and their derivatives have immunogenic potential which could again see a patient requiring immunosuppression.

Foetal liver cells and cord blood

Shafritz et al. (Oertel et al. 2007, Oertel et al. 2008, Oertel et al. 2006) have investigated the use of foetal hepatocytes in the treatment of adult hepatic diseases. They have demonstrated that healthy foetal hepatocytes can repopulate the normal liver over time without the requirement for selective depletion of the endogenous liver cells as is the case for adult transplantation in metabolic disease. Another foetal source of cells for liver therapy, cord blood has been suggested as a source of cells that can differentiate into hepatocytes (Newsome et al. 2003) however cord blood derived cells were actually fusing with resident hepatocytes (Fujino et al. 2007). Whilst it is tempting to consider cord blood as a suitable starting resource for adult stem cells (as it is regularly discarded after birth) it has to be concluded that the epithelial differentiation from cord blood occurs with very low efficiency, if at all.

Bone marrow derived stem cells

The ability of Bone Marrow (BM) stem cells to regenerate the haematopoietic system is well characterized. These cells are highly potent, and are able to reconstitute all lineages in the blood. Moreover initial data within the field suggested that the epithelial components of the adult liver could be replaced during injury and as such could offer a novel and importantly accessible source of liver epithelial cells (Alison et al. 2000, Petersen et al. 1999, Theise et al. 2000a,b). Recently however many groups have investigated this highly promising cell route and have concluded, using a variety of cell tracking techniques, that the contribution of the BM to epithelial regeneration in both experimental model systems as well as in gender mismatched human tissue from patients that have received organ transplantation is negligible, and cannot explain the initial observations made (Thorgeirsson and Grisham 2006, Vig et al. 2006). Furthermore in the FAH deficient mouse model, where large scale functional restitution of the liver had been achieved through bone marrow therapy (Lagasse et al. 2000) was in fact the result of cell fusion between BM derived macrophages and hepatocytes (Wang et al. 2003, Willenbring et al. 2004). These works together demonstrate an important relationship between the BM and the liver parenchyma, however it is suggested that the benefit garnered from this approach in a clinical setting may be insufficient to justify its widespread use.

Mesenchymal stem cells

Mescenchymal Stem Cells (MSCs) have been researched intensively recently because of their ease of availability and isolation, their ability to be expanded *in vitro* and apparent ability to differentiate into multiple cell types. MSCs that have been purified from BM, umbilical cord blood and adipose tissue have been differentiated into a "hepatocyte-like" cells (Banas et al. 2007, Hong et al. 2005, Lee et al. 2004). Aurich et al. (2009) has managed to identify and isolate MSCs from human adipose tissue and differentiate these cells into "hepatocyte-like" cells with appropriate hepatocytic functions. These cells engrafted into the liver of immunodeficient mice and subsequently contributed to parenchymal mass.

Cell therapy for liver cirrhosis

In both acute and metabolic disease the liver parenchyma is a receptive environment in which hepatocytes, and those cells which have been directed into hepatocytes can engraft, grow and contribute to the parenchymal mass. However in the context of late fibrosis and cirrhosis the liver is not receptive to these cells, and as such very little benefit can be anticipated from these cell types. The cirrhotic liver is a collection of apoptotic and inflammatory cells but also contains large regions of fibrotic matrix predominantly collagen and elastin which serves to partition the liver and results in the nodular pathology often observed in these diseases. In the context of hepatocyte transplantation many studies have attempted intra-splenic transplantation as a route of therapy, although this is clearly substantial, and may not sufficiently compensate for the large degree of parenchymal loss during cirrhosis, needless to say this does not address the secondary symptoms of disease such as cancer and portal hypertension.

Remodelling of the fibrotic and cirrhotic scar is crucial in the restoration of parenchymal architecture and function. Failure to do this results in the perpetuation of scar tissue, and loss of the metabolic parenchyma. Macrophages play a critical role in the restoration of liver architecture through the remodelling of this scar tissue. This has been critically examined using the DTR (Diphtheria Toxin Receptor) mouse model. The transgenic expression of human DTR allows the selective ablation of mouse macrophages using diphtheria toxin (Duffield et al. 2005). In this experimental scenario, depletion of macrophages during liver injury results in less rapid scar formation indicating that macrophages are important for the creation of liver fibrosis. During the recovery phase after liver injury, depletion of macrophages resulted in less reduction in the liver scar suggesting that macrophages also have a critical role in the remodelling process. We have shown that these macrophages are readily circulating from the bone marrow and as such are an important cell population to study with regard to disease modification (REF).

Rodent models of cell therapy for cirrhosis

Rodent models for cirrhosis are somewhat limited. In rats a cirrhotic liver with classical nodular regeneration is achieved through chronic and intermittent administration of carbon tetrachloride which induces hepatocyte necrosis, inflammation and scarring. This model works in a step like fashion, and exhibits the hallmarks of fibrosis progression and cirrhosis establishment. In the mouse severe fibrosis can be regularly seen using these regimes, however cirrhosis is rarely achieved.

Rodent studies for the treatment of liver cirrhosis using mesenchymal stem cells have given conflicting results in the literature (Russo et al. 2006, Abdel et al. 2007, Carvalho et al. 2008, Fang et al. 2004, Oyagi et al. 2006, Shi et al. 2005, Zhao et al. 2005). Early work indicates that down-regulation of liver fibrosis occurs in animals that have received MSCs as a therapy; however other groups have not been able to reproduce these results. These studies suggest that mesenchymal stem cells may not be ideal as starting cell populations for the treatment of liver cirrhosis. The MSC is the parent cell to fibroblasts, which are classically associated with matrix synthesis and deposition, in the context of disease these cells may be pro-fibrotic if infused into a cirrhotic liver, and as such may exacerbate the damaged and scarred phenotype observed in these livers. This underscores the more important principle that the long-term phenotype of infused cells must be well characterized.

Endothelial Progenitor Cells (EPCs) have also been shown to have antifibrotic and pro-regenerative potential. The investigators used single and repeated peripheral injections of *in vitro* cultured endothelial progenitor cells in a rat model of cirrhosis (Nakamura et al. 2007). These cells were able to reduce the amount of fibrosis in the recipient livers and also improve liver function tests. This was thought to occur by increased expression of certain matrix metalloproteinases (MMP-2, 9 and 13) and hepatocyte-like growth factor, vascular endothelial growth factor, transforming growth factor alpha and epidermal growth factor. It has been increasingly realized that functionally distinct subpopulations of macrophages exist within the liver during hepatic fibrosis and repair (Duffield et al. 2005). Furthermore, during the resolution of fibrosis, scar-associated macrophages express the collagenase MMP13 to mediate matrix degradation and remodel fibrotic liver (Fallowfield et al. 2007). Therefore, harnessing the appropriate populations of "anti-fibrotic" macrophages and using them in a therapeutic context seems a logical next step.

Human studies

The beneficial role of BM in the repair of diverse organs is increasingly appreciated (Korbling and Estrov 2003). As a result a wave of human studies intending to identify and define the role of BM cells in the restoration of normal liver function after disease ensued. It is now well accepted that the BM's physiologically relevant contribution in chronic liver disease is not through differentiation into parenchymal cells but rather supplying non-parenchymal cell types that may mediate the generation and resolution of fibrosis and support the liver's intrinsic regenerative compartments (Russo et al. 2006, Duffield et al. 2005). The literature as a whole now appreciates that both short-range, paracrine signalling to inflammatory mediators, as well as long-range cellular recruitment to the site of damage by these cells results in the restitution of fibrosis in the adult liver.

Cell types used in human studies

BM fractions can be enriched for HSCs on the basis of antigenic cell surface markers such as CD34 and CD133. Cell sorting increases the cost and complexity of the process; a potential solution is the use of less purified cell fractions. Mononuclear cells can be separated from BM relatively simply by density gradient centrifugation however, CD34+ cells constitute only 1–2% of this population (Steen et al. 1994) and the proportion of true stem cells within this fraction is even lower.

Granulocyte-Colony Stimulating Factor (G-CSF) is routinely used in clinical haematology to increase BM stem cell yields from peripheral blood which can be harvested though leukophoresis. Similarly, protocols for obtaining these cells from peripheral blood in the human studies of cirrhosis all used G-CSF to bolster stem cell numbers. This is an important step in therapy as it enables the patient to generate enough stem cells so that they might have repeat doses of cells during the time course of therapy. Moreover there is not a consistent improvement in liver function in cirrhotic patients given G-CSF alone (Lorenzini et al. 2008) indicating that both mobilization and reinfusion of these cells are critical for the positive phenotypes demonstrated in these studies.

Summary of clinical trials

• CD34+ cells

In a phase 1 clinical trial by Gordon and colleagues, CD34+ cells were isolated from peripheral blood after G-CSF administration and re-infused via the hepatic artery or portal vein (Gordon et al. 2006). Whilst not demonstrating a clear benefit in terms of liver biochemistry or assessed clinical parameters over the initial 2 mon period or on longer term follow up of 12–18 mon (Levicar et al. 2008), the primary endpoints of safety and acceptability were reached. The same group subsequently studied a series of nine patients with alcoholic cirrhosis who had been abstinent for at least 6 mon (Pai et al. 2008). CD34+ cells were isolated following G-CSF administration and leukapheresis and then expanded 5 fold *in vitro*. A mean dose of 2.3 $x10^8$ cells was injected into the hepatic artery. There were no complications and liver biochemistry improved over a 3 mon period.

In another uncontrolled study, four patients with decompensated cirrhosis were given 2.5–8 $x10^6$ BM derived CD34+ cells via the hepatic artery in a radiologically guided procedure (Mohamadnejad et al. 2007). One of these patients (autoimmune aetiology of cirrhosis) developed progressive renal failure and died of liver failure prompting early termination of the trial. The adverse chain of events was linked to radiocontrast nephropathy in a patient with advanced liver disease and pre-existing renal impairment however this highlights important issues regarding the route and delivery of cell therapy and also patient selection.

The most striking improvements in the literature were noted in two abstinent patients with decompensated alcoholic cirrhosis who received 4 or 6.93 $x10^6$ CD34+ cells/kg in a mononuclear cell fraction (Yannaki et al. 2006). These cells were obtained by leukapheresis following G-CSF mobilization and returned via a peripheral route. The clinical and biochemical recovery was maximal 1 yr after cell therapy.

• Unsorted mononuclear cells

A study of nine cirrhotic patients (3 HBV, 5 HCV, 1 unknown aetiology) from the Yamaguchi University Hospital in Japan given 2.21–8.05 $x10^9$ BM derived unsorted mononuclear cells via a peripheral vein showed significantly improved serum albumin and Child's-Pugh (CP) scores up to 6 mon later (Terai et al. 2006). Three of these patients also had paired liver biopsies which demonstrated increased immunohistochemical markers of regeneration 4 wk after therapy.

The largest published series to date of cell therapy for cirrhosis is from the University of Bahia in Brazil (Lyra et al. 2007). Ten patients with a variety of aetiologies (including HCV, alcohol and cholestasis) on the liver transplant waiting list received 1.6–13.1 $x10^8$ mononuclear BM cells via the hepatic artery. The treatment was safe over the 4 mon follow up period, including in a 70 yr old subject. There was no overall deterioration in blood tests or CP/MELD scores in these patients. Following this work, the same group presented the first randomized controlled data of cell therapy for cirrhosis. These data were published in abstract form in 2007 and outlined significant improvements in albumin and CP score against the control arm, using a similar methodology for cell selection and delivery (Lyra et al. 2007).

• MSCs

There is one report in the literature of MSC therapy in cirrhosis (Mohamadnejad et al. 2007). The group from Tehran University gave 1–6 $x10^7$ BM derived, *in vitro* cultured MSCs via a peripheral vein to four patients with decompensated cirrhosis from the transplant waiting list. One year later, the subjects were still alive with improved quality of life (on questionnaire); three of the four patients had a net reduction in MELD score.

Limitations of human work to date and future directions

Drawing these data together, cell therapy for cirrhosis seems to be safe when carried out in a carefully selected patient group and delivered in a controlled manner. The effects of cell therapy on liver disease are generally encouraging however, the lack of control groups in the published studies make it impossible to draw firm conclusions at this stage. The natural history of chronic liver disease tends to involve a degree of fluctuation often around a general downward trend. Notably in the context of alcoholic cirrhosis, liver function may markedly improve when the chronic insult is lifted. Therefore studies of relatively short duration without control arms are not sufficiently informative to prove a positive effect of cell therapy. The heterogeneity of chronic liver disease mandates carefully chosen inclusion criteria and also sufficient numbers of patients to adequately power such studies. Robust randomized, controlled trials are necessary to conform to the standards of evidence required for translation into a clinically acceptable therapy.

HSC populations have been shown to be safe in the work described here. The precise purity of the cells used will reflect a combination of practicality, cost and effectiveness. At present, our understanding of the mechanism of any effect is limited. Alternative therapeutic cell populations perhaps with more defined, predictable and beneficial phenotypes may add clarity to this area.

The chosen route of administration needs to balance any advantages of more local delivery of cell therapy against the ease and safety of using peripheral access. Variable numbers of cells have been given to patients both between and within studies. In future work, this must be standardized to accurately test the efficacy of cell therapy.

Currently, the measures of outcome in clinical trials of cell therapy are limited. The day to day variability of chronic liver disease means that narrow windows of data collection can miss important trends. Therefore it is important to collect serial data where possible to capture more subtle changes. This most easily lends itself to the key routine blood tests (such as prothrombin time and serum albumin, bilirubin and creatinine). Widely acknowledged shortfalls of the Childs-Pugh scoring system include the subjective assessment of encephalopathy and the vulnerability of albumin measurement and ascites estimation to recent events and interventions (e.g., infection/recent albumin infusion and diuretic usage/recent ascitic drainage respectively). The objectivity and validation of the MELD score in addition to its clinical familiarity offers several advantages. Liver biopsy has a recognized morbidity and mortality associated with it but can provide invaluable tissue for the analysis of effect and mechanism. The sampling error inherent in liver biopsy necessitates its interpretation in conjunction with other tests such as panels of serum markers of liver fibrosis (Rosenberg et al. 2004). Newer non-invasive tests such as indocyanine green clearance and ^{31}P magnetic reasonance spectroscopy are increasingly well recognized as representative biomarkers (Lim et al. 2003). The precise battery of measures used will depend on the cell type in question, its anticipated effect and the candidate patient population. A key area for technological development in the field of cell therapy is the tracking of donor cells. A topographical appreciation of applied cells *in vivo* will underpin mechanistic insights into cell localization and longevity.

In summary, the increasing understanding of the mechanisms of cell therapy emerging from animal work will continue to inform human trials. The next step in refining the clinical studies of cell therapy for chronic liver disease will involve well designed controlled studies able to test the important parameters determining translation to the clinic (Houlihan and Newsome 2008).

References

Abdel Aziz MT, HM Atta, S Mahfouz et al. 2007. Therapeutic potential of bone-marrow–derived mesenchymal stem cells on experimental liver fibrosis. Clinical Biochemistry. 40: 893–899.

Alison MR, R Poulsom, R Jeffery et al. 2000. Hepatocytes from non-hepatic adult stem cells. Nature. 406: 257.

Aurich H, M Sgodda, P Kaltwasser et al. 2009. Hepatocyte differentiation of mesenchymal stem cells from human adipose tissue *in vitro* promotes hepatic integration *in vivo*. Gut. 58(4): 570–581.

Banas A, T Teratani, Y Yamamoto et al. 2007. Adipose tissue-derived mesenchymal stem cells as a source of human hepatocytes. Hepatology Baltimore, Md. 46: 219–228.

Belardinelli MC, F Pereira, G Baldo et al. 2008. Adult derived mononuclear bone marrow cells improve survival in a model of acetaminophen-induced acute liver failure in rats. Toxicology. 247: 1–5.

Bilir BM, D Guinette, F Karrer et al. 2000. Hepatocyte transplantation in acute liver failure. Liver Transpl. 6: 32–40.

Carvalho AB, LF Quintanilha, JV Dias et al. 2008. Bone marrow multipotent mesenchymal stromal cells do not reduce fibrosis or improve function in a rat model of severe chronic liver injury. Stem Cells Dayton, Ohio. 26: 1307–1314.

Duffield JS, SJ Forbes, CM Constandinou et al. 2005. Selective depletion of macrophages reveals distinct, opposing roles during liver injury and repair. The Journal of Clinical Investigation. 115: 56–65.

Fallowfield JA, M Mizuno, TJ Kendall et al. 2007. Scar-associated macrophages are a major source of hepatic matrix metalloproteinase-13 and facilitate the resolution of murine hepatic fibrosis. J Immunol. 178: 5288–5295.

Fang B, M Shi, L Liao et al. 2004. Systemic infusion of FLK1(+) mesenchymal stem cells ameliorate carbon tetrachloride-induced liver fibrosis in mice. Transplantation. 78: 83–88.

Fisher RA and SC Strom. 2006. Human hepatocyte transplantation: worldwide results. Transplantation. 82: 441–449.

Fox IJ, JR Chowdhury, SS Kaufman and et al.1998. Treatment of the Crigler–Najjar syndrome type I with hepatocyte transplantation. The New England Journal of Medicine. 338: 1422–1426.

Fujino H, H Hiramatsu, A Tsuchiya et al. 2007. Human cord blood CD34+ cells develop into hepatocytes in the livers of NOD/SCID/gamma(c)null mice through cell fusion. Faseb J. 21: 3499–3510.

Gordon MY, N Levicar, M Pai et al. 2006. Characterization and clinical application of human CD34+ stem/progenitor cell populations mobilized into the blood by granulocyte colony-stimulating factor. Stem Cells Dayton, Ohio. 24: 1822–1830.

Hay DC, J Fletcher, C Payne and et al. 2008. Highly efficient differentiation of hESCs to functional hepatic endoderm requires Activin A and Wnt3a signaling. Proceedings of the National Academy of Sciences of the United States of America. 105: 12301–12306.

Hay DC, S Pernagallo, JJ Diaz-Mochon et al. 2011. Unbiased polymer library screening of polymer libraries to define novel substrates for functional hepatocytes with inducible drug metabolism. Stem Cell Research. 6(2): 92–102.

Hong SH, EJ Gang, JA Jeong et al. 2005. *In vitro* differentiation of human umbilical cord blood-derived mesenchymal stem cells into hepatocyte-like cells. Biochem Biophys Res Commun. 330: 1153–1161.

Houlihan DD and PN Newsome. 2008. Critical review of clinical trials of bone marrow stem cells in liver disease. Gastroenterology. 135: 438–450.

Kaji K, K Norrby, A Paca et al. 2009. Virus-free induction of pluripotency and subsequent excision of reprogramming factors. Nature.

Korbling M and Z Estrov. 2003. Adult stem cells for tissue repair—a new therapeutic concept? The New England Journal of Medicine. 349: 570–582.

Lagasse E, H Connors, M Al-Dhalimy et al. 2000. Purified hematopoietic stem cells can differentiate into hepatocytes *in vivo*. Nature medicine. 6: 1229–1234.

Lee KD, TK Kuo, J Whang-Peng et al. 2004. In vitro hepatic differentiation of human mesenchymal stem cells. Hepatology Baltimore, Md. 40: 1275–1284.

Levicar N, M Pai, NA Habib et al. 2008. Long-term clinical results of autologous infusion of mobilized adult bone-marrow–derived CD34+ cells in patients with chronic liver disease. Cell Proliferation 41 Suppl. 1: 115–125.

Lim AK, N Patel, G Hamilton et al. 2003. The relationship of *in vivo* 31P MR spectroscopy to histology in chronic hepatitis C. Hepatology Baltimore, Md. 37: 788–794.

Lorenzini S, A Isidori, L Catani et al. 2008. Stem cell mobilization and collection in patients with liver cirrhosis. Alimentary Pharmacology & Therapeutics. 27: 932–939.

Lyra AC, MB Soares, LF da Silva et al. 2007. A pilot randomised controlled study used to evaluate efficacy of autologous bone marrow mononuclear cells transplantation in patients with advanced chronic liver disease. Hepatology Baltimore, Md. 46: 1.

Lyra AC, MB Soares, LF da Silva et al. 2007. Feasibility and safety of autologous bone marrow mononuclear cell transplantation in patients with advanced chronic liver disease. World J Gastroenterol. 13: 1067–1073.

McKenzie TJ, JB Lillegard and SL Nyberg. 2008. Artificial and bioartificial liver support. Semin Liver Dis. 28: 210–217.

Medine CN, B Lucendo-Villarin, W Zhon et al. 2011. Robust generation of hepatocyte like cells from human embryonic stem cell population. Jove 56 ptii 2969 doi 10.3791/2969

Mohamadnejad M, K Alimoghaddam, M Mohyeddin-Bonab et al. 2007. Phase 1 trial of autologous bone marrow mesenchymal stem cell transplantation in patients with decompensated liver cirrhosis. Arch Iran Med. 10: 459–466.

Mohamadnejad M, M Namiri, M Bagheri et al. 2007. Phase 1 human trial of autologous bone-marrow–hematopoietic stem cell transplantation in patients with decompensated cirrhosis. World J Gastroenterol. 13: 3359–3363.

Muraca M, G Gerunda, D Neri et al. 2002. Hepatocyte transplantation as a treatment for glycogen storage disease type 1a. Lancet. 359: 317–318.

Nakagawa M, M Koyanagi, K Tanabe et al. 2008. Generation of induced pluripotent stem cells without Myc from mouse and human fibroblasts. Nature biotechnology. 26: 101–106.

Nakamura T, T Torimura, M Sakamoto et al. 2007. Significance and therapeutic potential of endothelial progenitor cell transplantation in a cirrhotic liver rat model. Gastroenterology. 133: 91–107 e101.

Newsome PN, I Johannessen, S Boyle et al. 2003. Human cord blood-derived cells can differentiate into hepatocytes in the mouse liver with no evidence of cellular fusion. Gastroenterology. 124: 1891–1900.

Oertel M, A Menthena, MD Dabeva et al. 2006. Cell competition leads to a high level of normal liver reconstitution by transplanted fetal liver stem/progenitor cells. Gastroenterology. 130: 507–520; quiz 590.

Oertel M, A Menthena, YQ Chen et al. 2006. Properties of cryopreserved fetal liver stem/ progenitor cells that exhibit long-term repopulation of the normal rat liver. Stem cells Dayton, Ohio. 24: 2244–2251.

Oertel M, A Menthena, YQ Chen et al. 2007. Comparison of hepatic properties and transplantation of Thy-1(+) and Thy-1(–) cells isolated from embryonic day 14 rat fetal liver. Hepatology Baltimore, Md. 46: 1236–1245.

Oertel M, A Menthena, YQ Chen et al. 2008. Purification of fetal liver stem/progenitor cells containing all the repopulation potential for normal adult rat liver. Gastroenterology. 134: 823–832.

Okita K, M Nakagawa, H Hyenjong et al. 2008. Generation of mouse induced pluripotent stem cells without viral vectors. Science. 322: 949–953.

Overturf K, M Al-Dhalimy, R Tanguay et al. 1996. Hepatocytes corrected by gene therapy are selected *in vivo* in a murine model of hereditary tyrosinaemia type I. Nature genetics. 12: 266–273.

Oyagi S, M Hirose, M Kojima et al. 2006. Therapeutic effect of transplanting HGF-treated bone marrow mesenchymal cells into CCl4-injured rats. Journal of Hepatology. 44: 742–74.

Pai M, D Zacharoulis, MN Milicevic et al. 2008. Autologous infusion of expanded mobilized adult bone-marrow–derived CD34+ cells into patients with alcoholic liver cirrhosis. Am J Gastroenterol. 103: 1952–1958.

Payne CM, K Samuel, A Pryde et al. 2011. Persistence of functional hepatocyte like cells in immune-compromised mice. Liver international. 31(2): 254–262.

Petersen BE, WC Bowen, KD Patrene et al. 1999. Bone marrow as a potential source of hepatic oval cells. Science. 284: 1168–1170.

Rosenberg WM, M Voelker, R Thiel et al. 2004. Serum markers detect the presence of liver fibrosis: a cohort study. Gastroenterology. 127: 1704–1713.

Russo FP, MR Alison, BW Bigger et al. 2006. The bone marrow functionally contributes to liver fibrosis. Gastroenterology. 130: 1807–1821.

Shi MX, BJ Fang, LM Liao et al. 2005. [Flk1+ mesenchymal stem cells ameliorate carbon tetrachloride-induced liver fibrosis in mice]. Sheng wu gong cheng xue bao = Chinese Journal of Biotechnology. 21: 396–401.

Si-Tayeb, FK Noto, M Nagaoka et al. 2010. Highly efficient generation of hepatocyte like cells from induced pluripotent stem cells. Hepatology 51(1): 297–305.

Slehria S, P Rajvanshi, Y Ito et al. 2002. Hepatic sinusoidal vasodilators improve transplanted cell engraftment and ameliorate microcirculatory perturbations in the liver. Hepatology Baltimore, Md. 35: 1320–1328.

Sokal EM, F Smets, A Bourgois et al. 2003. Hepatocyte transplantation in a 4-year-old girl with peroxisomal biogenesis disease: technique, safety, and metabolic follow-up. Transplantation. 76: 735–738.

Steen R, L Morkrid, GE Tjonnfjord et al. 1994. c-kit ligand combined with GM-CSF and/or IL-3 can expand CD34+ hematopoietic progenitor subsets for several weeks *in vitro*. Stem Cells. 12: 214–224.

Sullivan GS, DC Hay, IH Park et al. 2010. Generation of functional hepatic endoderm from human induced pluripotent stem cells. Hepatology 51(1): 329–335.

Takahashi K and S Yamanaka. 2006. Induction of pluripotent stem cells from mouse embryonic and adult fibroblast cultures by defined factors. Cell. 126: 663–676.

Takahashi K, K Tanabe, M Ohnuki et al. 2007. Induction of pluripotent stem cells from adult human fibroblasts by defined factors. Cell. 131: 861–872.

Terai S, T Ishikawa, K Omori et al. 2006. Improved liver function in patients with liver cirrhosis after autologous bone marrow cell infusion therapy. Stem Cells Dayton, Ohio. 24: 2292–2298.

Theise ND, M Nimmakayalu, R Gardner et al. 2000a. Liver from bone marrow in humans. Hepatology Baltimore, Md. 32: 11–16.

Theise ND, S Badve, R Saxena et al. 2000b. Derivation of hepatocytes from bone marrow cells in mice after radiation-induced myeloablation. Hepatology Baltimore, Md. 31: 235–240.

Thorgeirsson SS and JW Grisham. 2006. Hematopoietic cells as hepatocyte stem cells: a critical review of the evidence. Hepatology Baltimore, Md. 43: 2–8.

Vig P, FP Russo, RJ Edwards et al. 2006. The sources of parenchymal regeneration after chronic hepatocellular liver injury in mice. Hepatology Baltimore, Md. 43: 316–324.

Wang X, H Willenbring, Y Akkari et al. 2003. Cell fusion is the principal source of bone-marrow-derived hepatocytes. Nature. 422: 897–901.

Willenbring H, AS Bailey, M Foster et al. 2004. Myelomonocytic cells are sufficient for therapeutic cell fusion in liver. Nature medicine. 10: 744–748.

Yannaki E, A Anagnostopoulos, D Kapetanos et al. 2006. Lasting amelioration in the clinical course of decompensated alcoholic cirrhosis with boost infusions of mobilized peripheral blood stem cells. Exp Hematol. 34: 1583–1587.

Zhao DC, JX Lei, R Chen et al. 2005. Bone-marrow–derived mesenchymal stem cells protect against experimental liver fibrosis in rats. World J Gastroenterol. 11: 3431–3440.

11

Technical and Bioethical Challenges Associated with using Stem Cells for Research and Therapy

Jordi L.Tremoleda,[1,*] *Itziar de Lecuona*[2] and *Sian E. Harding*[3]

Introduction: development of cell-based therapies for liver diseases

Millions of patients worldwide suffer from end-stage liver disease. Orthotopic liver transplantation has rapidly advanced and is currently the treatment of choice for patient with end-stage liver disease. However, the procedure requires major surgery, with many liver transplant recipients needing to spend time in intensive care units in the post-operative period, with considerable risks for infectious complications, acute renal failure and/or poorly functioning grafts (Razonable et al. 2011). Given the donor shortage and that only one or two patients at most may benefit from one donor liver, and the complexity associated with the transplantation procedure various

[1]Investigator, MRC Centre for Clinical Sciences, Faculty of Medicine, Imperial College London, London W12 0NN, UK; Email: Jordi.lopez-tremoleda@imperial.ac.uk
[2]Lecturer, Dept. of Public Health, School of Medicine, Bioethics and Law Observatory, Barcelona Science Park, University of Barcelona, 08028 Barcelona, Spain;
Email: itziardelecuona@ub.edu
[3]Professor of Cardiac Pharmacology, National Heart and Lung Institute, Faculty of Medicine, Imperial College London, UK, London SW7 2AZ, UK, Member of the Nuffield Council on Bioethics; Email: sian.harding@imperial.ac.uk
*Corresponding author

alternatives have been evaluated, including cell therapies. The use of living cells as a therapeutic source to restore, maintain and/or enhance the liver function have numerous advantages when compared to organ transplantation as cells can be expanded *in vitro* to overcome the limits of organ shortage, cells can be genetically manipulated to correct functional and/or metabolic alterations, cells can be cryopreserved, transplanted without major surgical procedures and can be obtained from the same patients avoiding major risk of rejection and need for immunosuppressive treatments (Locke et al. 2009).Unfortunately adult hepatocytes cannot be expanded *in vitro* and cryopreserved cells are easily damaged during the freezing/thawing procedure

The transplantation of fresh isolated adult hepatocytes has been explored as an alternative to liver transplantation. Despite some encouraging results, demonstrating a clinical improvement for up to 12 mon following hepatocyte transplantation, this approach is hampered by the heterogeneity of patients treated, variety of transplantation reports with different setting and follow up studies (Sancho-Bru et al. 2009) and, importantly, the difficulties in harvesting and storing sufficient quantities of hepatocytes along with the significant cell loss following transplantation. All these factors have so far limited the potential applications of using adult hepatocytes for therapy (Han et al. 2009).

Alternative cell sources for hepatic cell therapy are being examined and of particular interest are both adult and foetal liver stem cells and pluripotent stem cells, with its great potential as an expandable and reliable cell source (Table 11.1). Stem cells are undifferentiated cells capable of proliferation, self-maintenance and are able to differentiate with plasticity into diverse mature progeny, including hepatocytes. Indeed, hepatocyte-like cultures have been generated *in vitro* from both Embryonic Stem Cells (ESCs), human peripheral blood monocytes and bone marrow-derived stem cells (Ruhnke et al. 2005, Agarwal et al. 2008, Chivu et al. 2009) and their administration in rodent models has been shown to support hepatic function (Sato et al. 2005, Moriya et al. 2008). Initial clinical pilot studies testing the direct administration of bone marrow derived stem cells have been encouraging, supporting and improving liver function in patients with chronic liver disease (Terai et al. 2006, Lyra et al. 2007). Indeed, Bone Marrow Stem Cells (BMSCs) have long been recognized as possessing potential to support hepatic population. A mobilization of the bone marrow derived hematopoietic stem cells fraction has been observed during hepatic injury and seems to play an important role in hepatic regeneration (Russo et al. 2006). However, the bone marrow derived mesenchymal stem cell fraction can potentially contribute to liver fibrosis (Forbes et al. 2004), highlighting the complexity of the injury/regeneration process in the liver. Further revisions on the role of bone marrow-derived hepatocytes in preclinical and clinical studies have indeed highlighted remarkable differences in the way that cells may support this cellular regenerative process and this is likely to have an important impact in the development of therapeutic approaches (Stutchfield et al. 2010). Two main cellular regenerative processes to injured liver have emerged including a) a direct contribution to the resident hepatocyte population and progenitor cells and b) the supportive indirect role of bone marrow derived stem cells to promote endogenous processes. In the latter, investigations for therapies are largely focussed on investigating the paracrine mechanism by which bone marrow stem cells may promote tissue repair and how to mobilize the endogenous cell and/or paracrine factors resources in the patients.

In this chapter we will address the different technical and bioethical challenges associated with the development of stem cell treatments for hepatic disease. The main sources of stem cells that have been proposed for cell transplantation are described and the practical challenges for their use as models of human disease and their potential for clinical applications are discussed, with a particular emphasis on the use of induced reprogrammed pluripotent stem cells.

Sources of stem cells for cell therapy

Embryonic stem cells

ESCs are pluripotent cells derived from the inner cell mass of the blastocyst-stage embryos and possess potent differentiation potential as they can generate any differentiated phenotype of the three primary germ layers (endoderm, mesoderm and ectoderm), as well as germ cells (Thompson et al. 1998). Moreover, due to their capacity for self-renewal they can theoretically provide an unlimited supply of cells that could be differentiated into hepatocytes to support regeneration of the diseased liver. *In vitro* differentiation of ESCs towards the hepatic lineage is well documented, generating functional but immature hepatocytes (Yamamoto et al. 2003, Agarwal et al. 2008) and when they are transplanted in animal models of hepatic disease, these ESCs-derived hepatocytes were able to engraft in the damaged liver and support differentiation towards hepatocytes but with limited regenerative and function capacity (Heo et al. 2006). Ongoing studies are focussed on improving the differentiation protocols to generate more robust hepatocyte-like cells from ESCs with greater functional and regenerative properties (Hay et al. 2008, 2011, Payne et al. 2011).

However, bridging the therapeutic potential of human Embryonic Stem Cells (hESCs) towards its clinical applications has raised one of the most controversial debated areas in scientific research. This debate largely revolves around the ethical implications of using human embryos as the main source for obtaining these promising stem cells, mostly from surplus embryos created for the purpose of assisted reproduction. Such controversy also derives from the potential application of somatic cell nuclear transfer to generate blastocysts in order to produce immunologically compatible hESC lines for therapeutic use in patients. The most controversial argument against the research with human embryonic stem cells is that this technique involves the use of human embryos and, as largely argued, this represents the destruction of human life for those who considers life begins at conception (Green 2007). Similarly, the creation of embryos specifically for the sole purpose of deriving stem cells may have more considerable moral implications than using surplus embryos that would eventually be destroyed. The main point of argument is that there are two approaches to assess the embryo, the biological one, by which that embryo is part of the development process as proven scientific consensus and the metaphysic-theological approach which debates the issue strictly in terms of absolute moral values. In this vein, any regulatory system should be based on scientific facts and integrating the real facts that affect society, not in specific moral beliefs (Casado and Egozcue 2000).

Opponents to the use of embryonic stem cells cite the advantages of adult stem cells and, more recently, induced pluripotent stem cells. The moral objection to ESCs has had the effect of driving forward the research in these alternative areas more strongly. On the one hand, this can be beneficial in developing new lines of enquiry, but it may also have the consequence of accepting a lower standard of human material.

Furthermore translational research with human ESCs has also raised other controversies related to the mixing of human and animal cells or DNA. Studies using chimaeras are common in biomedical research and introduction of human DNA into animal cell lines (or vice versa) is an everyday occurrence. Vaccines and xenotransplantation research represent some of the more visible examples in which animal-human mixing has been routinely applied. Similarly, human ES cells are implanted into immunodeficient mice to test for teratoma formation as the standard to assess stem cell quality and developmental potential (Lensch et al. 2007). Indeed, the International Society for Stem Cell Research (ISSCR) endorse the use of these forms of human/animal chimeras on which a limited number of human cells are introduced at any stage of pre-or post-natal development, and where incorporation into any lineage or tissue is likely to be minimal. However, those protocols in which human cells may contribute a significant degree of chimerism to the central nervous system and/or germ line raised serious ethical concerns as prospects that they may develop human features. This is especially concerns when human ES cells might be incorporated into the brain or gonads of a closely related primate (Hyun et al. 2007). The National Academy of Sciences (NAS) and the ISSCR have formulated guidelines for research involving human-animal chimeras and recommend that "all research involving the introduction of human ES cells into nonhuman animals at any stage of pre/post-natal development should be reviewed by the Embryonic Stem Cell Research Oversight committees" (ESCRO; NRC-U.S. 2005, NAS 2007) and stated their opposition to research in which human ES cells are introduced into non-human primate blastocysts (pre-implantation embryos), as well as the breeding of any animal into which human embryonic ES cells have been introduced.

Another aspect associated to the clinical use of human ESC-derived therapies into patients is the potential risk of teratoma formation and the immunocompatibility issues following transplantation. Several preclinical studies are addressing the safety issues related to phenotype stability of the derived ESCs but the real proof remains to be validated through long-term trials (Wu et al. 2007). Similarly, there are concerns about the compatibility between ESCs-derived tissues between patients. Even though ESCs seem to display a certain degree of immune privilege due to their minor expression of histocompatibility antigens, it has been shown to be sufficient to induce an acute rejection in differentiated tissues derived from ESCs (Robertson et al. 2007). Several approaches have been proposed to overcome the immune barrier including the use of somatic cell nuclear transfer from given donors to generate blastocysts and their derived patient-specific hESCs lines that are immunologically compatible for therapeutic use. However, as mentioned there are also serious ethical concerns with the use nuclear transfer technique in human embryos and the targeted "creation" of human embryos for the sole purpose of deriving hESCs.

Professional groups including the NAS in the US, the Human Fertilisation and Embryology Authority within the UK and the ISSCR clearly oppose to reproductively clone humans and prohibit *in vitro* culture of human embryos beyond 14 d. These guidelines allow the derivation of human ESCs lines from excess embryos from IVF, from embryos created explicitly for human ESC research, or from embryos created by therapeutic cloning, including the necessity for appropriately detailed and informed consent. Such approval would be only granted where the research is "necessary or desirable" and the use of human embryos is essential. It is of interest that these guidelines also include the possibility for interspecies mixing, in particular regarding the production of *cybrids*, in which human somatic cell nuclei are introduced into enucleated animal oocytes to induce reprogramming. Such procedures have been recently approved in the UK but only after a considerable debate (St John and Lovell-Badge 2007). While the use of this technology may overcome the difficulties of using human oocytes to carry out similar procedures, mostly compromised by the shortage of donated human oocytes (Holden 2005), there are important scientific concerns mostly associated with the presence of both animal and human mitochondria (St John and Schatten 2004). The overall idea is that stem cells obtained from human-nonhuman cybrid embryos could be suitable for *in vitro* purposes of studying human diseases, despite the presence of animal mitochondria (St John and Lovell-Badge 2007). However, these cells may not be fully functional after transplantation *in vivo* into animal models as mitochondrial function would be essential in this setting. Similar to nuclear transfer, the complexity and poor understanding of the intrinsic mechanisms along with the limited results have critically compromised the further progression of this technique.

Adult stem/progenitor liver cells and bone marrow stem cells

The postnatal liver has an extensive regenerative potential grossly supported by the presence of a progenitor cell population and the plasticity of the hepatocyte. In response to an injury and/or regenerative stimulus, the normally quiescent hepatocytes will become activated and initially drive the regeneration response. However, such regenerative response of resident hepatocytes may be overwhelmed and/or compromised and then the Hepatic Progenitor Cell population (HPCs) will support the hepatic regeneration, as detected in several hepatopathies (Zhou et al. 2007). However, the identification and further characterization of this putative liver progenitor cell population remains controversial and warrants further research. To add to the complexity, the presence of such cell populations in the liver has been related to the existence of multiple stem cell niches within the liver that can become activated depending on the mechanisms and location of injury (Petersen and Shupe 2008). In rodents these liver stem cells have been termed oval cells and are capable of differentiation into both hepatocytes and biliary epithelia. In humans, the presence of a similar stem cell niche population in the adult liver and their role in hepatic regeneration remains uncertain. Despite all these hurdles, the potential application of these adult stem/progenitor cells remains interesting as it avoids many of the ethical issues related to ESCs and safety issues associated to teratoma formation, allowing the development of autologous transplantation with no need for immunosupression. Ongoing studies are focussing on the characterization of these liver stem cells and their functional regenerative role during liver disease.

Although liver regeneration is mainly an endogenous process, the supportive role of extra-hepatic Bone Marrow Stem Cells (BMSC), in particular the HSCs subpopulation, that migrate into the liver and contribute to its regeneration has long been recognized (Petersen et al. 1999, Forbes et al. 2004). Several hypotheses about the mechanism by which BMSC contribute to liver regeneration have included differentiation into hepatocytes, cell fusion with hepatocytes to form cell hybrids and stimulation of paracrine effects (Stutchfield et al. 2010). Although transdifferentiation of progenitor cells into hepatocytes and the differentiation of hepatocyte like cells have been demonstrated, the mechanism that seems to be more favourable clinically is the role of BMSCs in modulating the endogenous repair mechanisms through their paracrine effects promoting angiogenesis which is crucial to liver regeneration (Beaudry et al. 2007).

HPCs and/or BMSCs have relevant benefits as cell sources for autologous transplantation in liver disease. However, there are some critical issues associated with, their identification and characterization, to obtain a well-standardized homogeneous cell population, and further concerns over the phenotypic stability of the engrafted BMSCs, their risk for contributing to liver fibrosis and whether the regenerative potential of cells derived from diseased patients may already be critically compromised. Despite all, autologous stem cells derived from bone marrow are the only stem cell type to have undergone clinical investigation to date. However, published studies have been small and mostly reporting safety, with limited information on efficacy of repair (Piscaglia et al. 2010). Further randomized controlled trials are needed to establish a genuine role of these cells in liver repair.

Alternatives approaches to deriving pluripotent stem cell lines

There is a general agreement on the great potential for human health in cellular-based therapies. Through recent years we have seen extensive debates discussing alternative sources to obtain pluripotent cells, with the aim to reinforce the development of these cell-based therapies and to bring new resources that may help to somehow soften the stringent attitude overshadowed by all the debate on the use of human embryos as a source (Fig. 11.1).

One proposed method is a **modified approach for the nuclear transfer technique**, in which a single blastomere from an eight to 16-cell embryo is used to create a hESC line while not compromising the potential of the embryo to develop further. This has been successfully carried out in mice (Chung et al. 2006) and it is also clinically used for Preimplantation Genetic Diagnosis (PGD). The isolated cell can be cultured, expanded and used for genetic diagnosis and the derivation of embryonic stem cells line. With more than a decade of experience in this technology and hundreds of children born following PGD, the evidence suggests that the procedure does not impair the embryo's developmental capacity with no proven risk for higher malformations or related developmental problems in the born children (Verlinsky et al. 2004). Moreover, researchers have reported the successful development of pluripotent stem cells lines from single cells taken from

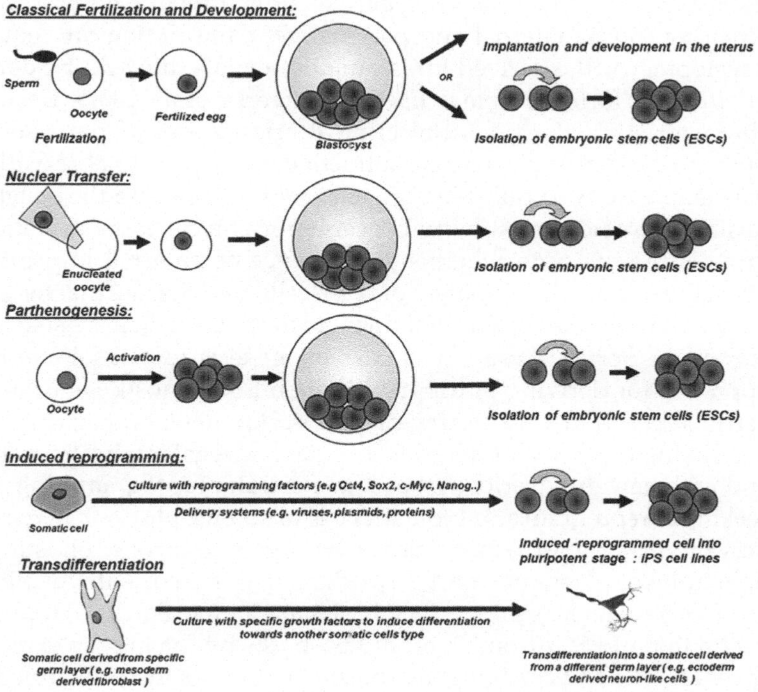

Figure 11.1. Approaches to generating pluripotent stem cells: Derivation of human embryonic stem cells from the inner cell mass of blastocysts that form after fertilization of the oocyte by spermatozoa, after somatic-cell nuclear transfer into an enucleated oocyte or by direct activation of the oocyte (parthenogenesis). Alternatively, pluripotent stem cells can be obtained by direct reprogramming of somatic cells by culture with factors such as Oct4, c-Myc, Sox2, Nanog (induced pluripotent stem cells: iPSCs). ESCs: embryonic stem cells.

thawed human embryos (Klimanskaya et al. 2006), providing the proof-of-concept for the feasibility of this approach. However, it is not clear that this could be implemented as a routine approach for couples undergoing ART procedures, as they would more likely opt for implanting an embryo that has not been exposed to such manipulations. Although evidence suggests the removal of a single blastomere imposes very little risk on a child that is born as a result of this procedure, there is much uncertainty about whether such an approach could become a standard procedure for isolating and developing personalized pluripotent stem cell lines from hESCs.

Another potential source of pluripotent embryonic stem cells is through **parthenogenesis**, which involves the development of an embryo directly from an oocyte without fertilization by sperm. Parthenogenetic oocyte activation can be induced in the absence of spermatozoa by exposure to certain chemicals (e.g., ionomycin, cycloheximide) and physical stimuli (e.g., electric stimulation) in various mammalian species including humans

(Revazova et al. 2007). Although parthenote embryos do not produce viable offspring (both maternal and paternal gene imprinting are required for development), the derived ESCs can differentiate into cellular derivatives *in vitro* and form teratomas *in vivo* (Lengerke et al. 2007). Despite such characteristics, the potential clinical application of parthenogenetic human ESCs has not been very attractive mostly due to observations that parthenote-derived cells are not genetically identical to the oocyte donor raising immunocompatibility concerns for clinical transplantation. Also there is the question whether the absence of paternal imprinting will affect the normal development of these cells, and the fact that the source of oocytes are very restricted and limited to the patients that are able to donate an oocyte. For all these issues parthenogenesis remains a controversial approach for deriving pluripotent human ESCs, mostly associated to its "artificiality" and their lacking of full developmental capacity.

Taking into account the ethical, social and legal implications involved, and with new discoveries in the area of **cellular reprogramming** research, seeking to reprogram adult somatic cells to become pluripotent represents a very promising approach and hence the large amount of public support and high number of research groups working in this field. With the pioneering work of Yamanaka's group (Takahashi and Yamanaka 2006) reporting the generation of ESCs from an adult fibroblasts by forcing the expression of specific genes and changing the epigenetic status of the adult differentiated cell to become a pluripotent cell, has brought a whole new perspective in the stem cell field. Undoubtedly, this opens whole new perspectives for continuing the development of stem-cell derived therapies, as such technology might obviate the need for destroying the embryo and have great potentially facilitating the derivation of an immune compatible cell-based therapeutic products from patients, overcoming the need for immunosuppressive treatments. However, all the gene manipulation work also raises some other safety and bioethical concerns regarding the epigenetic stability of the differentiated cell state.

Finally another approach of remarkable growing interest is the induction of **transdifferentiation** of somatic cells into other differentiated lineages. Indeed, this approach argues whether cells can be induced to "transdifferentiate" directly into another state of differentiation, converting from a somatic cell type to another type without first reprogramming into pluripotent cells (Graf and Enver 2009). This transdifferentiaton between somatic cell lines of cells it is well reported in several studies, with examples of conversion within the same germinal layer such as fibroblast into muscle cells (Weintraub et al. 1989) within the hematopoietic lineage converting T and B-cells into macrophages (Xie et al. 2004) and cardiac fibroblasts into cardiomyocytes-like cells (Ieda et al. 2010). Recent studies have reported the transdifferentiation potential between different germ layers, Vierbuchen et al. (2010) converted fibroblast (mesoderm) to neurons-like cells (ectoderm) and Efe et al. (2011) converted mouse fibroblast to cardiomyocytes, shedding some new light on the biology underlying cell differentiation and cell-fate, suggesting new avenues for induced reprogramming protocols obviating the need to pass through a pluripotent state.

Overall, this approach for inducing differentiation is increasingly being investigated and ongoing research is providing good evidence of cell plasticity; however it also brings with it the risk of aberrant gene expression patterns and therefore serious consequences for its clinical use.

Prospects of iPSCs for disease modelling and cell transplantation

Recently, the generation of Hepatocyte-Like Cells (HLCs) has also been demonstrated to be feasible with human-induced pluripotent stem cells (iPSCs) (Song et al. 2009, Sullivan et al. 2010, Touboul et al. 2010, Si-Tayeb et al. 2010). IPSCs appear to be a promising source for the generation of hepatocyte-like cells that could provide a defined and renewable source of human cells relevant for cell therapies and pharmacological *in vitro* testing. But there are several technical challenges regarding the reprogramming procedure and the expansion and derivation of pluripotent-induced cells and their differentiation towards functional hepatocytes. These issues will be discussed below.

Introduction to cell reprogramming and induction of pluripotency: technical issues

Cells are characterized by their gene expression patterns and function. For a cell to achieve its differentiated status during development a series of changes which take place in a tightly regulated manner to allow the adequate modifications in the genetic profile and function. To investigate these epigenetic changes is crucial to understand how cell fates are regulated and thus how can this be controlled physically.

During development, cells proceed from a state of totipotency, pluripotency to a more differentiated and tissue restricted fate. As the embryo develops, forming the trophoblast lineage and the inner cell mass, cells are characterized as pluripotent because they can be differentiated into all somatic cells and germ line cells of the developing embryo. Such changes through the cell potency to differentiate into specific lineages depend on tightly regulated intrinsic molecular signalling pathways. However, cell fates during development are neither restrictive nor irreversible. Initial studies in cell reprogramming, using nuclear transplantation of somatic nuclei into eggs (Gurdon 1962) showed that the epigenome of differentiated cells can be reset to a pluripotent state. Thus to elucidate the factors that maintain the pluripotent state of the early embryonic cells and their modulation during differentiation is the key to understand the potential application of reprogramming cell technology in basic and translational research.

Reprogramming of somatic cells to pluripotency is accompanied by extensive remodelling and epigenetic modifications. Despite the research much remains to be understood about the nature of gene regulation during iPSCs formation. Although several strategies are used to induce reprogramming, only a small fraction of the initial cells will become iPSCs, with the first reprogrammed cells appearing no earlier than 5–10 d after expression of the reprogramming factors (Jaenisch and Young 2008). The determination of efficiency of *in vitro* reprogramming is typically based upon the arbitrary appearance of iPSCs colonies and being calculated by the cells that expressed pluripotency markers (Hanna et al. 2010). This methodology, while informative, remains purely descriptive and provides very limited information on intrinsic cell changes and their expansion and differentiation potential. Moreover, it does not account for differences in cell population size, cell division times, fluctuations in the number of cells that are lost or undergoing apoptosis during passaging and cell culture raises variability among the cell population undergoing reprogramming.

Many studies suggest that somatic cells can be reprogrammed effectively to a pluripotent state with molecular and biological characteristics indistinguishable to those for ESCs. However, it is important to realize that accurate and quantifiable assessment of these characteristics remains challenging mostly due to the methodological limitations and the complexity of the dynamics of the cellular reprogramming process. Some of the main limitations that may affect the gene expression of the iPSC and their biological characteristics are the presence and incomplete silencing of the reprogramming transgene used to induce the reprogramming the cells of origin (Kim et al. 2010, Sullivan et al. 2010a). These residual transgenes can perturb the identity and functionality of the induced cells. Another factor is that the *in vivo* developmental competency of the iPSCS may also depend on the original genetic background of the cells of origin, and they may respond differently to the induced reprogramming. Other critical parameters relate to the effects of the expanding protocols as reprogrammed cells may adapt differently to culture conditions. Overall these constraints may affect the epigenetic state and biology of the iPSCs which can translate into serious genetic deregulation events and abnormal developmental potential.

One of the main issues regarding the potential use of iPSCs in the clinics as a good alternative to hESCs is how similar these pluripotent reprogrammed cells are to ESCs. Are iPSCs equivalent to ESCs? Have the iPSCs the same differentiation potency as their embryonic counterparts? Can the genetic and/or epigenetic profile of the reprogrammed cells be affected or destabilized during differentiation to targeted lineages and/or during transplantation, generating cells that are different from those derived from ESCs? All these questions raise important biological issues regarding the safety and the efficacious use of this cell source for developing cell-based therapies, and also its ethical and legal implications.

Biological assays and gene expression studies are the main criteria generally used to assess the quality of the iPSCs. The chimera formation and germ line contributions assays by which iPSCs derived cells prove capable of forming all cell types of the body are crucial to prove that the iPSCs are pluripotent. In fact in mice it has been proven that the iPSCs have the same developmental potential as ESCs (Zhao et al. 2009). However, these types of assays are impossible in the human system (Lensch et al. 2007, Dolgin 2010). In this field most tests rely on *in vitro* pluripotency assays and teratoma formation. Indeed, the teratoma assay is currently the only established means of demonstrating how human iPSCs possess pluripotency when placed in an *in vivo* system. However, this approach remains a qualitative test and it is difficult to have a quantitative comparative approach with this type of assay. Other tests relying on identifying the genetic signature of these cells allow for a more comparative approach between iPSCs and ESCs. Several studies have indicated the similarities between these cells but the gene expression signatures remain controversial, in particular in early passage iPSCs lines in which larger variation in chromatin structure and gene expression are observed (largely associated with the residual epigenetic memory of the cell of origin (Polo et al. 2010)), and this may affect their differentiation potential (Sullivan et al. 2010b). The fact that such patterns are not seen in ESCs derived from nuclear transfer suggests that reprogramming *in vitro* with transcription factors may be suboptimal.

Several studies have reported differences in gene expression, patterns of DNA methylation and differentiation potential (Chin et al. 2009, Doi et al. 2009, Hu et al. 2010). Many of these studies have focussed on few cell lines and therefore it is difficult to systematically study the role of epigenetic and transcriptional variation. In a recent study researchers tested a large panel of 16 iPSCs lines derived from multiple donors of varying age, sex and health status and examined their pluripotency and their ability to generate terminally differentiated cells, in particular, motor neurons (Boulting et al. 2011). Most of the iPSCs were capable of expressing similar pluripotent markers and generating functional neurons under a stringent standard protocol in a very efficient manner. These procedures were highly reproducible between laboratories, indicating the robustness of the standardized protocols as quality control for this stem cell resource. This extensive study found that human iPSCs could be differentiated on average as well as hESCs, supporting their similarities. Although there were some variations in the differentiation efficiency from individual human iPSCs, these differences were highly reproducible suggesting that this may be related to intrinsic characteristics of the cell lines and not their reprogramming.

Interestingly, in a recent study Sullivan et al. (2010a), showed that iPSCs lines derived from both female and male sexes and from two different ethnicities (North-American Indian and Caucasian) could be successfully differentiated to hepatocytes-like cells at efficiency similar to that for human ESCs (iPSCs efficiency of 70–90% vs. hESCs efficiency of 90–95%). It was suggested that iPSCs may prove a more uniform starting cell pool for derivation hence the better response to the differentiation protocols but iPSC-derived hepatocytes seem to be less functional that those derived from hESCs. Further research is required to clarify these possible differences in susceptibility to differentiation and function between pluripotent cell lines. Such studies are the key to the progression of these iPSCs derived cells towards modelling and clinical use.

Applications for modeling human diseases

Human pluripotent stem cells have the potential to generate all tissues of the body (Thomson et al. 1998, Park et al. 2008) which present exciting opportunities for *in vitro* modelling of specific human diseases. They have great potential for investigating pathogenesis, aiding therapeutic discovery and exploring functional genomics (Colman and Dreesen 2009, Freund and Mummery 2009), in particular for those disorders without suitable animal models and/or those previously lacking lineage specific cells for *in vitro* studies (Fig. 11.2).

It is hoped that human iPSC-derived cells can provide complementary information even for diseases in which animal models are available. The limitations of animal models are being realized, especially by pharmaceutical companies who have experienced high attrition rates when compounds are transferred from animal experiments to human. The initial approach evolved from researchers isolating cells from preimplantation embryos used for genetic diagnosis purposes and deriving "disease-specific" human embryonic stem cells from the embryos affected by some genetic disorders such as cystic fibrosis and Huntington's disease (Mateizel et al. 2006). This allowed for the generation of mutant hESC lines that could be used as disease models. In this vein, the development of pluripotent stem cells derived from reprogrammed adult somatic cells harvested from patients with specific disorders provides new scope for creating disease-specific cell lines for modelling (Saha and Jaenisch 2009). These techniques have been applied successfully to blood and skin derived cells and their potential to

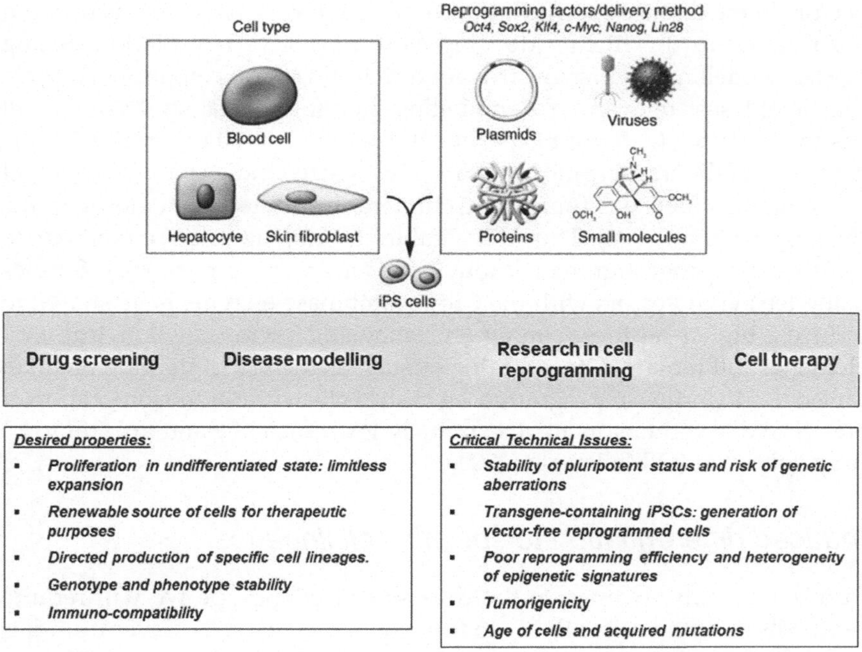

Figure 11.2. Schematic representation of the generation process of iPSCs and their research and clinical applications: different tools can be used to reprogram different adult cells inducing it to become an "undifferentiated" and to regain pluripotency; A number of desirable factors influencing the properties of induced pluripotent stem cells and constraints that may affect their biology and potential clinical use.

progress towards a disease phenotype has been demonstrated (Ebert et al. 2009, Raya et al. 2009).

We will now address the practicalities and critical issues associated with the development of cell sources for modelling from human iPSCs.

Harvesting cells from a patient and induction of pluripotency

Somatic cells are harvested from patients; this is usually done by tissue biopsy or blood sampling (e.g., a cell fraction expressing the surface protein CD34 from the general blood sample or fibroblasts from skin samples). Then reprogramming protocols will be implemented to generate cell lines that will be critically screened for their pluripotent phenotype.

The direct reprogramming strategies currently available are using viral vectors generating human iPSCS with multiple integrated copies of these viral transgenes. The possibility of persistence of these integrated reprogramming vectors in the disease specific human derived iPSCs poses the risk of being reactivated during further development and/or differentiation during disease modelling leading towards abnormal and/or cancerous phenotypes. This is of lesser relevance for modelling than for clinical studies, but still has the potential to disrupt experimental results. A well reported example is the use of the transcription factor c-Myc which studies have shown that led to high incidence of tumours in chimeras generated from mouse iPSCs (Nakagawa et al. 2008). Thus it is vital to develop alternative methods to induce cell reprogramming without using viral vectors and c-Myc. Studies using lentiviral vectors with the Cre-recombinase enzyme have shown to induce a higher reprogramming efficiency and vector deletion, but viral elements still remain. Other strategies such as the use of peptide or small molecules to induce reprogramming in the cells without the integration of any viral derived factors have extremely low efficiency and are still to be fully validated (O'Malley et al. 2009).

Induced differentiation to specific cell lineages

In order to fully investigate the disease cell phenotype we will require protocols to differentiate iPSCs to functional somatic cells. Several studies have successfully reported the differentiation of HiPSCs into different cell lineages (neuronal: Chambers et al. 2009, hepatocytes: Sullivan et al. 2010a-see Fig 11.3; fibroblasts: Hokemeyer et al. 2008). However, such protocols are not well standardized, with different time courses and varying effectiveness between lines. Differentiated cells will undergo an extensive functional and biochemical analysis to identify their phenotype and how its matches the specific phenotype associated to the disease pathology. Due to the complexity of multigenetic disorders, this approach is mostly valuable for those monogenic disorders with well characterized phenotype disorder patterns (Raya et al. 2009, Ebert et al. 2009). The application of this approach for disorders with more complex genetic phenotypes associated to multifactorial disorders remains rather difficult. Indeed, disease associated to complex and significant epigenetic modifications pose a serious challenge for modelling. It is clear that obtaining a disease related phenotype from a given patient will critically depend on the characterization of the disease phenotype and its comparison with healthy control induced pluripotent cells. In addition, most derived cells will display immature phenotypes compared to authentic adult cells. Certain aspects of pathology are likely only to be observed in the adult phenotype. To complicate matters, development of disease can be associated with regression to a foetal phenotype, for example in heart, making it difficult to distinguish between immaturity and disease-related characteristics.

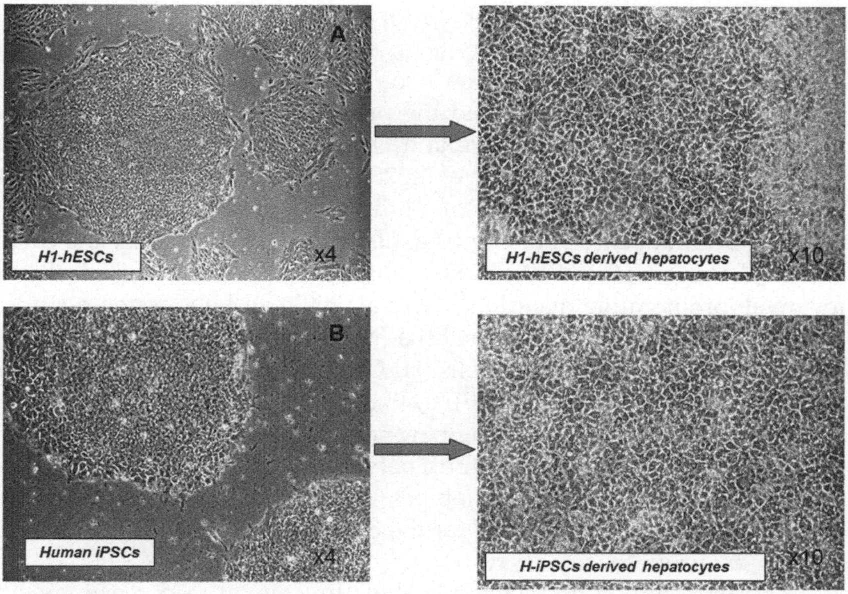

Phase Contrast microscopy images of hepatocyte-like cells derived from hESCs (A) and hiPSCs (B). Images provided by Dr Claire N. Medine (MRC CRM, University of Edinburgh)

Figure 11.3. Phase Contrast microscopy images of hepatocytes derived from H1-hESCs (A) and hiPSCs human iPSCs cells. Images kindly provided by Dr Claire N. Medine (MRC CRM, University of Edinburgh).

Generation of disease-relevant phenotypes from the pluripotent iPSCs

The main objective is to ensure that efficient differentiation methods are implemented to induce the differentiation of the pluripotent iPSCs derived from specific patients towards the disease-relevant phenotype (Fig. 11.2). The use of reporter genes is extensively applied to identify genetic modifications during cell differentiation to specific phenotypes, which may help to understand the dynamics of genetic changes associated to specific progression towards a disease phenotype. However, the integration of these reporter genes can also be quite inefficient for human ESCs, with limited transduction efficiency (Xia et al. 2007). Nevertheless, this is a growing area of research and new techniques are evolving to facilitate the incorporation of reporter genes for targetting endogenous genes in human iPSCS (Hockemeyer et al. 2009).

Remarkable progress has already been made with some cases of modelling monogenic diseases and some complex genetic disorders with early or short term developmental pathogenesis. Park et al. (2008) reported the production of human iPSC lines for 10 diseases, ranging from simple Mendelian inheritance disorders (e.g., Adenosine deaminase deficiency) to complex conditions (e.g., Parkinson's disease and Type 1 Diabetes). Other groups have also reported the production of iPSC lines for other neurological, myeloproliferative and endocrine disorders (Dimos et al. 2008, Lee G et al. 2009, Soldner et al. 2009, Raya et al. 2009, Maehr et al. 2009). Importantly, the number of studies on human iPSCs in which the disease-specific *in vitro* phenotype has also been reported is also increasing (Carvajal-Vergara et al. 2010). There are also existing human iPSCs cell lines for myeloproliferative disorders (Ye et al. 2009) and endocrine disorders such as the juvenile diabetes mellitus (Maehr et al. 2009), and this list is growing fast. Hepatocyte-like cells (HLCs) have also been generated from human IPSCs (Sullivan et al. 2010a, Inamura et al. 2011) opening whole new perspectives for developing new sources of hepatocytes from iPSCs derived from patients suffering form polymorphic metabolic disorders and other liver associated disease genotypes (e.g., alpha-1-antitrypsin), to model liver disorders *in vitro* allowing for the development of novel biomarkers and *in vitro* drug toxicity assays.

Another challenging factor is that the relevant –disease specific phenotype may also require specific environmental stimuli, from the neighbouring cells and surrounding tissues, cell matrix and mechano-physical properties. Therefore we need to provide an *in vitro* environment that will mimic this extracellular milieu during co culture of the differentiated disease-specific human iPSCs. Extensive work is ongoing in the field of biomaterial engineering to develop powerful supportive co culture *in vitro* systems to provide effective rich context for studying disease related cell-cell interactions and extracellular matrix effects (Guilak et al. 2009). The use of iPSCs derivatives with several bioactive materials and other evolving 2D and 3D scaffolds (Hay et al. 2011) may open new perspectives for cell culture, studying differentiation and disease modelling. A particular example for the liver is the development of micro patterned cell clusters from human hepatic cells to study liver function and hepatotoxicity assays (Khetani and Bhatia 2008). Recently Spence et al. (2011) efficiently directed the differentiation of human IPSCs into a three dimensional functional intestine which has great potential for generating intestinal tissue *ex vivo*.

Nevertheless, there are still some limitations when modelling human disorders, the latent period associated with some disorders such as Alzheimer's or late onset Parkinson's disease are especially difficult to mimic. Therefore for some diseases, an *in vivo* approach may be more suitable. Chimera assays provide a long-term access to complex and changing environment context for studying iPSCs; however there are some ethical concerns associated with this experimental approach. Human-animal interspecies chimeras can be generated by grafting pluripotent cells into embryos, fetuses or adult animals (see Behringer 2007). The transplantation of human cells into immunocompromised animal models is extensively used in regenerative research. Similarly, humanized mouse models are being developed with successful hematopoietic, neural and hepatic reconstitution with human stem cell derivatives (Friese et al. 2006). In particular, a hepatocyte-humanized mouse has been generated for exhibition of human-type responses for drug-related cytotoxicity assays and studying the pathogenesis of some viral infections (Kneteman and Mercer 2005).

Perspectives for clinical applications of IPSCs: Bioethical, legal and social issues

Despite the rapid progress in understanding cellular reprogramming to induce pluripotency, many technical and biological hurdles remain before their translation into clinical applications. Summing up the main challenges already discussed, including the inefficient direct reprogramming methodology, the difficulties for expanding the induced pluripotent cells in a robust and stable manner and the standardization of the molecular profiling on the cells, it is obvious that researchers need to establish reliable and reproducible standardized protocols to test pluripotent cell lines in the laboratory. Despite extensive research efforts focussed in setting up reliable test set for these cells (Boulting et al. 2011), a recent study reported aberrant epigenomic reprogramming in five human iPSCs cells, showing significant reprogramming variability and aberrant DNA methylation profiles (Lister et al. 2011). Hence the importance of ensuring that extensive and thorough studies are carried out to characterize the genomic signature of the reprogrammed cells for their safe and effective use in the clinics. One would envisage the clinical applications of the iPSC as follows: a) harvesting of somatic cells taken from a living patient or a frozen tissue bank; b) generation of pluripotent cells carrying a disease-linked mutation by using standard well-characterized reprogramming protocols, and as already discussed, preferentially using methods that do not leave any genetic footprint in the derived cells (e.g., using non integrating vectors, excisable genetic elements, or using chemical or protein induced methods; Saha and Jaenisch 2009, Kim et al. 2009, Hanna et al. 2010); c) differentiation of the induced pluripotent cells into the target cell type of choice using robust differentiation protocols for the purpose of cell therapeutic source and/or as supportive cell source for the development of bioactive artificial devices. These differentiated cells should be thoroughly characterized for their genetic signature and phenotype as a quality test for their differentiation; d) investigation of their potential therapeutic interest through *in vitro* and preclinical evaluation studies; e) long term safety studies to ensure that the cell source is safe and efficacious for human clinical use. The timescale of this process may pose a problem for the treatment of rapidly developing conditions.

Despite the general belief that the use of iPSCs raise fewer ethical concerns than those related to embryonic stem cells it must be acknowledged that, as reprogramming technology evolves, there are important dilemmas that have to be addressed (Fig. 11.4). These divide into donor-related issues such as the consent to donate material for iPSC derivation, and to matters of the safety of these untested therapies as they are translated towards clinical trials. It is crucial to realize the potential scaling prospect of the iPSCs derivatives and to understand whether the original donor will be

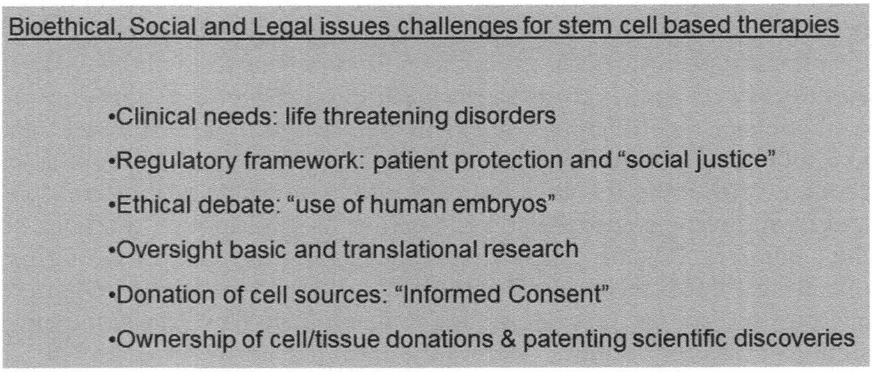

Figure 11.4. Factors influencing stem cell based therapies towards the clinics.

able to foresee a clear picture of their potential downstream use. These cells may be used extensively by researchers who will carry out a large variety of studies worldwide including genetic modification of the cells, preclinical testing in animal models, large scale genome sequencing, and sharing cell lines with other researchers. This raises issues of confidentiality protection and also as therapeutic applications evolve further issues with patenting, intellectual property and commercialization of the cell derivatives with/without proprietary right and share of royalties (Lo and Parham 2009). Several ethical, legal and social aspects must be considered for a responsible transition of stem cell research into appropriate clinical applications, as discussed in the Guidelines approved by the ISSCR (Hyun et al. 2008).

It may also be envisaged that selecting which disorders should be targeted for such experimental therapies is likely to raise a substantial bioethical debate. Clearly this is going to be a political decision that could be supported by legislation promoting specific areas of research. However, such legal response should also be supported by a social debate, where the scientific background, with its expectations and limitations, are clearly discussed. It is important to not generate false expectations or misconceptions in patients affected by the disease being studied.

Oversight of iPSC research and translation

First of all, it is imperative that individuals with stem-cell-specific expertise assist in the assessment of translational research, supporting an independent and thorough oversight of the studies that will lead to the clinical trials. This will involve a good scientific evaluation of the *in vitro* and *in vivo* preclinical studies that will form the basis for proceeding towards clinical application. This is particularly relevant when assessing potential treatment with iPSCs as there are technicalities that must be addressed as already discussed. Approved principles of quality control must be in place to promote maximal quality and safety of the cells to be used, and all is supervised by expert and independent oversight committee. It is important to acknowledge the complexities associated with some cell types and to ensure a systematic assessment of the integrity and potency of the cell products to minimize any risk to patients. Preclinical testing plays a key role in assessing the safety and establishing a proof of principle for therapeutic effect. Good preclinical studies must support the clinical strategy, providing convincing evidence of the safety of using these cells before advancing to human studies. Some of these studies can be particularly challenging as some human disease conditions may not be easily translatable to animal models and also there may be some existing physiological differences between species. Another challenge is that stem cells can act through multiple mechanisms and thus predicting their behaviour may prove difficult. Indeed, this is of particular interest for the iPSCs, in which it is crucial to prove that the reprogramming method worked well. The major stem cell banks such as the NIH's Human Pluripotent Stem Cell Registry requires the researchers submitting new cell lines of teratoma formation as it is currently considered the gold standard for assessing pluripotency. However, there is increasing debate on the efficacy and reliability of this test (Dolgin 2010). Some researchers argue that there may be no need to categorize for pluripotency as long as the cells can differentiate towards the required cell phenotype, e.g., hepatocytes for liver diseases. But others maintain that is crucial to ensure that that the cell line's developmental capacity is fully tested to ensure a safe and effective derivation of differentiated cells. There is the risk that if we attempt to shortcut the characterization of targeted derived cells from iPSCs we may miss important developmental pathways that are key for the "normal" differentiation of the cells towards a mature phenotype. Moreover, the *in vivo* teratoma assay has the added value that cells are injected into a living animal, thereby taking advantage of all the in situ growth factors that help to develop three-dimensional structurized tissues. Nevertheless, extensive ongoing work is focused on the development of genomic and bioinformatics' methods as alternative methods for appraising a cell line's pluripotency (Müller et al. 2010). However, until further consensus is achieved among the stem cells community, quantitative *in vitro* assays may so far remain as a good alternative in development.

These arguments demonstrated the complexity for developing reliable translational assays hence the importance of ensuring that individuals with stem cell expertise are involved in reviewing such translational research and to ensure that only when compelling preclinical data are available is there a justification for moving into clinical trials.

Consent for the derivation of human iPSCs

As with all clinical research, when planning for clinical trials of stem cell-based interventions the internationally accepted principles governing ethical conduct of clinical research and the protection of human subjects must be guaranteed. This requirement is identified as a fundamental international standard by the Helsinki Declaration (WMA 2004), the Council for International Organization of Medical Sciences (CIOMS 2002), the National Commission for the protection of Human Subjects of Biomedical and Behavioral Research (The Belmont Report 1979) and the ISSCR among others (ISSCR 2006, US-FDA 2007). From the international legal perspective the Convention on Human Rights and Biomedicine, 1997 "The Oviedo Convention" and its Additional Protocol on Biomedical Research, 2004 (Council of Europe) and the Universal Declaration on Bioethics and Human Rights, UNESCO, 2005 should be taken into account.

Undoubtedly many of the research uses with iPSCs may evoke concerns about privacy and confidentiality. Donors may not be fully aware of the whole research potential and they might consider certain aspects a violation of their privacy: for example, in large scale genomic studies one could foresee complexity with DNA databases and confidentiality issues (Lowrance and Collins 2007). That is why it is crucial to give consent specifically about the future consequences; it is important to realize that to revoke the consent may not imply that the information generated until then will all be deleted. (Seoane et al. 2008). However, the Council of Europe permits research on identifiable bodily material without consent (but subjected to ethical review) in specific occasions where the research addresses important scientific interest, there is no evidence for an expressed opposed consent and it is not possible within reasonable efforts to contact the person for the consent.

In the initial stages of translating these iPCS-derived cell based therapies, we would envisage that the donation would be for allogeneic use towards the development of patient-specific iPSCs derived therapies. In this case it may be useful that along with the voluntary informed consent, the donor is provided with all the information related to all the preparation and health screening of the cells, their storage and its duration and that the donor may be approached in the future to seek additional consent for new users. A relevant issue is the protection of donor privacy and the confidentiality of the acquired information, and the terms for their disclosure. This can raise some concerns on the future use of any resulting cell-derived products and the ownership of commercial and intellectual property rights. In some countries the regulatory bodies specifically mentioned the gratuity of the tissue/organ donation as a principle.

A key requirement for recipients of stem cells and cell products is also voluntary informed consent. This is particularly endorsed by the ISSCR guidelines and also by many of the jurisdictions regulating stem cell research in most of the westernized countries (Caufield et al. 2007, ISSCR Guidelines 2008, McGuire et al. 2008). Special emphasis needs to be placed on the unique risk of stem cell-based clinical research during the informed consent process, including cell proliferation, phenotype stability and/or tumour development, exposure to animal source materials, risk associated with viral vectors and possibly other unknown risks. Indeed, the informed consent is particularly challenging for these stem-cell therapies, and thus patients must be told about the realistic potential for therapeutic benefits as, the consent must emphasize the novel and experimental aspects of these cell based interventions. Information related to risk and benefits are crucial to avoid misconception on the therapeutic outcomes. It should be noted that for an adult who lacks capacity to make a decision regarding his/her body material, participation in research is only lawful if the research has the capacity to benefit the person, or where the risk involved is negligible. Hence the importance of providing detailed recommendations for patient selection when seeking appropriate consent for future research.

Property rights associated to iPSCs and derived products

Indeed, the use of personal genetic information by third parties has important ethical implications, mostly associated to the commercial use of these cells or the disclosure of personal information that could lead to discrimination towards the donor (e.g., disclosure of genetic predisposition could compromise employment and fairness assessment by insurance companies). To avoid such discrimination due to genetic causes and/or personal health, and to ensure the protection of sensitive personal medical information, the regulatory bodies encourage a strong level of protection and respect to such personal data as clearly stated in the article 10 of the Private life and right to information of the Oviedo's Convention (Council of Europe, Convention on Human Rights and Biomedicine 1997) and its

Additional Protocol on Biomedical Research (2004). Such guidelines and regulations tend to protect the privacy and confidentiality of the donor, avoiding any genetic or any kind of discrimination based on data obtained. While one could argue that the rational use of these derived cells should be only used for the benefit of the donor, it is important to ensure the benefits of such research are fairly and justly shared. In this vein, another key principle to consider is the benefit to other patients with similar disorders, not only to those directly involved in the donation and clinical trials; in agreement with the social justice values stated in the Universal Declaration on Bioethics and Human Rights by the UNESCO (2005).

Bodily material collected in the course of health interventions and/or collected and used for research purposes, with the consent of the patient concerned, may be stored in cell banks for long term. As the collected material (e.g., cells, tissue biopsies) have a unique identity, as they may be derived from the body of a person, this raises important bioethical concerns regarding the ownership and commercialization of this bodily material. Generally there can be no property rights in a human body, living or dead. The rights of individuals to their own bodies are not legally of "property ownership", as individuals cannot be owned as property by others. However, it is well established that where body parts "have acquired different attributes by virtue of the application of skill" then they may become property. For example any form of tissue that has been processed into a product, including modified tissues or cells, may be considered property and may legitimately be commercialized by those who have undertaken the procedure and not the person who originally donated the source of material (Rv Kelly and Lindsay 1998). Indeed, while The Human Tissue Act (UK 2004) prohibits any commercialism of human material for transplantation, this does not cover any parts that may have acquired the character of property by virtue of modification/application of human skill.

Regarding the issue of commercialization and financial gain, The Oviedo Convention and Additional Protocols stated that "the human body and its parts shall not, as such, give rise to financial gain"; as also stated in the Convention on Human Rights and Biomedicine (Chapter VII-prohibition of financial gain and disposal of a part of the human body). However, as these body parts are "processed" they become property and potentially a lucrative product for sale and resale (European Group on Ethics in Science and new technologies, European Commission 2012) creating a whole new perspective regarding the ethical, legal and financial aspects of commercialization of these "processed" products.

As stem cells technologies move forward, property and financial issues are becoming increasingly relevant, and it is up to the regulatory bodies and the oversight ethical committees to ensure that these are addressed with the appropriate proportionality, protecting the donor's rights and providing all the adequate information during the informed consent process but also protecting the research community and the industry interest for continuing advancing and investing in such technologies.

Sensitive research with iPSCs

Although there are several areas of research with iPS cells that raised some socio-ethical concerns, the fields of transplantation into humans and reproductive manipulation remain particularly controversial (Aalto-Setälä et al. 2009).

Despite the technical issues associated with reprogramming, we can foresee that "virus-free reprogrammed" –iPSCs will develop in the near future making human transplantation possible. In addition to the development of allogeneic transplantation applications, it is very likely that with the development of stem cell banks access to a large number of iPSCs lines may be feasible. As previously discussed, this may raise concern on the control that a donor may be able to exert on the iPSCs derived and expanded from their own cells; some people may not want their cells to be transplanted into another person, or may want to restrict the use of these cells for particular areas of research and we should ensure that the donor's autonomy is thoroughly respected.

Therefore it is important that all the aspects regarding the "ownership" of bodily parts that have been manipulated are clearly reflected in the consent procedure and that during the informed consent procedure the donor is provided with sufficient information about the possibility for transplantation. Even in the case of autologous transplantation, it should be explained that the procedure may be long and thus regular screening of the derived cells through this period will be critical.

Furthermore, it is should be recognized that during this procedure the donor conditions may worsen towards a terminal state of disease or a life-threatening injury. In such conditions, it is very likely that end-of-life decisions may affect the potential use of the donors' stem cells and/or their derivatives. The merits of ensuring that that some kind of advance care directives have been previously discussed and agreed with the donor include the respect of the donor's autonomy in decision making, the respect to his/her personal values and the complexity of involving third persons such as family or health care providers. Advance care directives allow patients to provide instructions about their preferences regarding the care they would like to receive if they develop a terminal illness or a life-threatening condition. However, these directives cannot predict what situation may arise in the future or what new therapies may be available, which is particularly critical for stem cell derived treatment.

Critically ill and/or end-of-life patients may not want to go ahead with the stem cell treatment at that stage. A clear example is patients undergoing hematopoietic stem cell transplantation which is a high risk procedure that is performed with a curative intent in patients with certain hematologic malignancies, but it is associated with significant short-term morbidity and mortality. To confront the potential loss of decision-making capacity by advanced patients, it is important to encourage them to engage in advance care discussions (Joffe et al. 2007). Interestingly, some clinical studies have shown the relevance of confronting such discussions with the patients, indicating that the lack of engagement in advance care decision is associated with increased mortality suggesting that patients may well benefit from considering these issues (Ganti et al. 2007). The process of incorporating stem cell derived therapies with advance care planning may be difficult and complex as many of these techniques are still in development, especially for the iPS cell technologies.

Another sensitive area of research is the differentiation of stem cells into primordial germ cells that then can differentiate into mature gametes. (Ko and Scholer 2006). Although gametes derived from iPSCs would be useful for understanding the process of gametogenesis and towards the development of therapeutic approaches for infertility, their use also raises serious ethical concerns as they could potentially lead to the creation of another human being. Moreover, gametes derived from iPS cells would virtually have the same DNA as the somatic cell isolated from a specific donor. Indeed, there is a critical moral responsibility when attempting reproduction for the creation of embryos, since it raises all the ethical aspects related to the moral status of the embryo and its human entity (Green 2007). Indeed, such type of reproductive research while ethically arguable by many as a relevant experimental approach for infertility treatments would also create important social objections to the iPSC technology. At the early stages of the iPSCs technology the field should try to restrict this type of research through explicit regulations and appropriate consent from donors.

Social justice considerations

Another potential issue is related to subject selection, ensuring fair access to well-designed clinical trials and effective stem-cell therapies without regards to patients' financial status, insurance coverage, social background or beliefs. It is important that stem cell research endorse the fairness of the social benefit of its progress. The ISSCR guidelines recommend that researchers and regulatory bodies attend to these issues of social justice and fairness when addressing the preparation of clinical trials and to ensure that benefits of research and discoveries are justly shared (Hyun et al. 2008). In this vein, the sponsor and the investigators have an ethical responsibility to make good efforts to secure sufficient funding to ensure that no eligible candidates are prevented from enrolling to any trial because of their inability to cover the costs of experimental treatment. Groups or individuals must not be excluded from the opportunity to participate in clinical stem cell research without rational justification.

Despite all these guidelines and the efforts of regulatory bodies, there are growing concerns that the development of customized cells and cellular products will remain unavailable to the vast majority of the population, mostly for financial reasons. While it is difficult to predict the future approach of the biotechnology industry and national health services regarding investment on these customized stem cell based therapies, at these early stages of translational development it is crucial that all the research community pool their efforts to facilitate international collaborations and universal access to stem cell-based treatment. These efforts may enable the implementation of more efficacious and cost-effective derivation and manufacturing procedures facilitating the establishment of large-scale banks for the benefit of the population.

Regulatory issues relevant to cell therapy

A broad variety of stem- cell derived products are likely to be developed for therapeutic purposes, including simple biomaterials and autologous cells through pluripotent, viral modified cells with full spectrum of risk for the patients. Moreover, stem cell therapies are likely to be developed through very different routes than pharmaceuticals and therefore guidelines for safety and efficacy testing of pharmaceuticals and medical devices may not be suitable for regulating such stem cell-based products. Therefore to ensure that these are regulated in a proportionate, fair and adequate manner, several regulatory agencies are working together to establish an effective legislation. In particular, it is essential that countries work together to promoting good standards and importantly, a flexible approach facilitating the development of new treatments for the benefit of patients. It is therefore necessary to develop a regulatory framework capable to strike the balance between fostering research for the general benefit and the protection of the rights, well-being and security of those involved. Again, this would also be an active role of the oversight ethics and research committees as previously discussed.

Stem cell therapy is considered one of the Advanced Therapy Products (ATP), together with gene therapy and tissue engineered products. A regulatory framework is required for these ATP's to ensure patient accessibility to products and governmental assistance for their regulation and control. The guideline has to be multidisciplinary and address the development, manufacture and quality control, as well as the preclinical and clinical development.

Legislation on cell therapy in ***Europe*** is based in three Directives: Directive 2003/63/EC (amending Directive 2001/83/EC), which defines cell therapy product as clinical products and specifies their requirements; The Clinical Trials Directive 2001/20/EC, which lays down the rules for conducting clinical trials to establish the safety, efficacy and quality of medicine products approved in the EU; and The Human Tissues and Cells Directive 2004/23/EC which establishes the standard quality, safety for the donation, harvesting, processing, preservation, storage and distribution of human tissues and cells. In addition, different levels of risk can be associated with specific types of stem cells, for example, the risk profile associated with induced pluripotent stem cells is expected to be different from those of adult stem cells for which substantial amount of clinical experience has already been gained. Therefore, a specific risk based approach according to Annex I, part IV of Dir 2001/83/EC can be applied to such pluripotent stem cells containing medicinal products. These technical requirements stated in this Annex are included in the Regulation 9EC) (No) 1394/2007 on advanced therapy medicinal products which addresses general aspects of manipulated and genetically modified cells. The marketing authorization in Europe has to be prepared through the European Medicines Agency (EMA) which provides several guidance documents on cell-based medicinal products (Guidelines on human cell-based medicinal products EMEA/CHMP/410869/2006) which cover the general aspects of all cell-based products including stem cell advanced therapy medicinal products. In particular for stem cells that are genetically modified, of particular interest for iPSCs, see the draft for the guidelines on the quality, preclinical and clinical aspects of medicinal products containing genetically modified cells (EMEA/CHMP/GTWP/671639/2010). Recently the European Medicines Agency released a new document on drug products that are manufactured using stem cells (EMA/CAT/571134/2009). Titled, Reflection Paper on Stem Cell-Based Medicinal Products, the document advises manufacturers on quality-control issues regarding these products and provides an overview of the use of stem cells in drug development. The document applies directly to those companies pursuing marketing authorization for stem cell-based products.

In the USA, the Food and Drug Administration (FDA) has jurisdiction over the production and marketing of any stem cell-based therapy involving transplantation of human cells into patients. The FDA provides the regulatory structure regarding human cells, tissues and cellular and tissue based products, covering the wide range of stem cell-based products that may be developed for therapeutic purposes. This is subjected to the Public Health Act, Section 361, which sets the regulatory framework that prevents the use of contaminated tissues or cells, limits the improper handling of tissues and ensures the clinical safety and efficacy of cells or tissues that are highly processed, are used for other than normal function, are combined with non-tissue components, or are used for metabolic purposes (FDA 1997). Stem cell-based products are also subjected to the Public Health Act, Section 351, which regulates the licensing of biological products to be used in studies involving humans (Halme and Kessler, 2006). This regulation envisions most, if not all, stem cell-based therapies to be considered biological products, and the manufacturer must ensure that product is "safe, pure and potent" (42 USCA § 262); the application for a new proposed stem cell-based product must provide data from preclinical studies on the likely safety and efficacy of the investigated product; and that approval and license of the product must be granted by the FDA once sufficient data demonstrates that the investigational product is safe and effective in humans. The key points of the current FDA regulation for cell therapy products include a) demonstration of clinical safety and efficacy; b) no risk for donors of transmission of infectious or genetic diseases; c) no risk for recipients of contamination or other adverse effects of cells or sample processing; d) specific and detailed determination of the type of cells forming the product and what are their exact purity and potency; e) *in vivo* safety and efficacy of the product. Moreover, specific recommendations are included regarding the use of cells or tissue that have been manipulated and may pose greater risk of disease transmission. Standardized procedures for processing and testing are required for the derivation, expansion, manipulation, banking and characterization of stem-cell products (21 C.F.R. § 610.12.). There have been concerns also about the potential use of stem cell products derived in xenogeneic feeder cells, and the FDA has established specific testing for adventitious agents according to the guidelines for xenotransplantation (U.S. Public Health Service Guideline on Infectious Disease Issues in Xenotransplantation). Another area which is becoming increasingly relevant relates to the safety concerns of cells that have undergone genetic alterations. The possibility that induced pluripotent cells could acquire genomic alterations that make them prone to transformation must be considered. New technologies are evolving to efficiently investigate genomic alterations and abnormalities. Recent studies have shown that karyotypically normal hiPSC lines possess small deletions and duplications that can be detectable by high profile genomic screening (Chin et al. 2009). Although this screening process is fast improving and new reprogramming techniques using non-integrative techniques are being developed, it is very likely that an IPSC line that has to be expanded through culture for a long time will have some genomic changes. This is challenging for the FDA when addressing the monitoring of iPSC lines; it is very likely that their approach for approving the clinical use of these cells may require a diligent monitoring of genomic alterations combined with all the other issues associated with transplantation (Lowry and Quan 2010).

There is still much work to be done and lots to be learned about the procedures to establish the safety and efficacy of iPSC based therapy products. The regulatory framework is very likely to evolve, as main regulatory bodies are actively trying to work in close collaboration with industry and academia to provide appropriate structure for ensuring the safety and efficacy of this new generation of stem cell-base products. Although at this stage the regulations may permit the use of xenobiotic reagents (based on their level of safety) and even viral integrative method for reprogramming cells (all under GMP grade approval), some effects of genomic instability or mutagenesis might not arise until several years of treatment. Indeed there are important hurdles to be addressed and such collaborations with the regulatory bodies are crucial towards the generation of safe iPSCs-based therapies.

Conclusion

Despite all the technical, ethical, social and legal hurdles associated with the induced reprogramming technology and the derivation of iPSCs, research in the field is rapidly advancing toward the clinic. IPSCs hold enormous potential for fundamental stem cell research, disease modelling and clinical therapies. The generation of isogenic cells by direct genetic reprogramming with well defined factors through a standard and well defined protocol seems quite challenging, but exciting times to come are just around the corner. Recently, the field has taken an important step forward by giving the green light for two stem cell trials, one using cells derived from hESCs for the treatment of spinal cord injury and the other using adult stem cells that have been genetically engineered to be conditionally immortal for the treatment of ischemic stroke (Mack 2011). While it is very hard to predict what will happen with any of these therapies, it is very good to see that the fields are taking such steps forward. From this perspective, the prospects for implementing the iPS cell technology in disease modelling and patient specific cell therapies in the near future are very bright. And surely it will demand a multidisciplinary cooperation involving fundamental research, preclinical testing, clinical translation, bioethical considerations and economical policies.

It is important to address all the technical issues and challenges that are discussed to promote a responsible translation of stem cell research into safe and clinically effective stem cell- based therapies for various human disorders.

References

Aalto-Setälä K, BR Conklin and B Lo. 2009. Obtaining consent for future research with induced pluripotent cells: opportunities and challenges. PLoS Biol. 7: e42.

Accessed on 4 March 2011.

Agarwal S, KL Holton and R Lanza. 2008. Efficient differentiation of functional hepatocytes from human embryonic stem cells. Stem Cells. 26: 1117–1127.

Beaudry P, Y Hida, T Udagawa et al. 2007. Endothelial progenitor cells contribute to accelerated liver regeneration. J Pediatr Surg. 42: 1190–1198.

Behringer RR. 2007. Human-animal chimeras in biomedical research. Cell Stem Cell. 1: 259–262.

Boulting GL, E Kiskinis, GF Croft et al. 2011. A functionally characterized test set of human induced pluripotent stem cells. Nat Biotechnol. 2011. 29: 279–286

Carvajal-Vergara X, A Sevilla, SL D'Souza et al. 2010. Patient-specific induced pluripotent stem-cell-derived models of LEOPARD syndrome. Nature. 465: 808–812.

Casado M and J Egozcue. 2000. Declaration of the Opinion Group of the Bioethics and Law Observatory on Embryo Research. M Casado and J Egozcue [eds.]. Signo. Barcelona, Spain.

Caulfield T, U Ogbogu, E Nelson et al. 2007. Stem cell research ethics: consensus statement on emerging issues. J Obstet Gynaecol Can. 29: 843–848.

Chambers SM, CA Fasano, EP Papapetrou et al. 2009. Highly efficient neural conversion of human ES and iPS cells by dual inhibition of SMAD signaling. Nat Biotechnol. 27: 275–280.

Chin MH, MJ Mason, W Xie et al. 2009. Induced pluripotent stem cells and embryonic stem cells are distinguished by gene expression signatures. Cell Stem Cell. 5: 111–123.

Chivu M, SO Dima, CI Stancu et al. 2009. *In vitro* hepatic differentiation of human bone marrow mesenchymal stem cells under differential exposure to liver-specific factors. Transl Res. 154: 122–132.

Chung Y, I Klimanskaya, S Becker et al. 2006. Embryonic and extraembryonic stem cell lines derived from single mouse blastomeres. Nature. 439: 216–219.

Colman A and O Dreesen. 2009. Pluripotent stem cells and disease modeling. Cell Stem Cell. 5: 244–247.

Convention for the Protection of Human Rights and Dignity of the Human Being with regard to the Application of Biology and Medicine: Convention on Human Rights and Biomedicine (Oviedo, 1997). Available http: //conventions.coe.int/Treaty/en/Treaties/html/164.htm Accessed on 4 March 2011.

Council for International Organizations of Medical Sciences. 2002. CIOMS international ethical guidelines for biomedical research involving human subjects, revised draft. Available http: //www.cioms.ch/publications/layout_guide2002.pdf. Accessed 4 March 2011.

Dimos JT, KT Rodolfa, KK Niakan et al. 2008. Induced pluripotent stem cells generated from patients with ALS can be differentiated into motor neurons. Science. 321: 1218–1221.

Doi A, IH Park, B Wen et al. 2009. Differential methylation of tissue- and cancer-specific CpG island shores distinguishes human induced pluripotent stem cells, embryonic stem cells and fibroblasts. Nat Genet. 41: 1350–1353.

Dolgin E. 2010. Putting stem cells to the test. Nat Med. 16: 1354–1357.

Draft Annex I, part IV of DIR 2001/83/EC applied to Advanced Therapy Medicinal Products. 2009. Available http: //www.ema.europa.eu/docs/en_GB/document_library/Scientific_guideline/2010/01/WC500069264.pdf Accessed on 4 March 2011.

Ebert AD, J Yu, FF Rose et al. 2009. Induced pluripotent stem cells from a spinal muscular atrophy patient. Nature. 457: 277–280.

Efe JA, S Hilcove, J Kim et al. 2011. Conversion of mouse fibroblasts into cardiomyocytes using a direct reprogramming strategy. Nat Cell Biol. 13: 215–222.

Forbes SJ, FP Russo, V Rey et al. 2004. A significant proportion of myofibroblasts are of bone marrow origin in human liver fibrosis. Gastroenterology. 126: 955–963.
Freund C and CL Mummery. 2009. Prospects for pluripotent stem cell-derived cardiomyocytes in cardiac cell therapy and as disease models. J Cell Biochem. 107: 592–599.
Friese MA, LT Jensen, N Willcox et al. 2006. Humanized mouse models for organ-specific autoimmune diseases. Curr Opin Immunol. 18: 704–709.
Ganti AK, SJ Lee, JM Vose et al. 2007. Outcomes after hematopoietic stem-cell transplantation for hematologic malignancies in patients with or without advance care planning. J Clin Oncol. 25: 5643–5648.
Graf T and T Enver. 2009. Forcing cells to change lineages. Nature. 462: 587–594.
Green RM. 2007. Can we develop ethically universal embryonic stem-cell lines? Nat Rev Genet. 8: 480–485.
Guidelines on human cell-based medicinal products EMEA/CHMP/410869/2006. Available http: //www.ema.europa.eu/docs/en_GB/document_library/Scientific_guideline/2009/09/WC500003898.pdf Accessed on 4 March 2011.
Guidelines on quality, non-clinical and clinical aspects of medicinal products containing genetically modified cells. EMEA/CHMP/GTWP/671639/2010. Available http: //www.emea.europa.eu/docs/en_GB/document_library/Scientific_guideline/2010/06/WC500093309.pdf Accessed on 4 March 2011.
Guilak F, DM Cohen, BT Estes et al. 2009. Control of stem cell fate by physical interactions with the extracellular matrix. Cell Stem Cell. 5: 17–26.
Gurdon JB. 1962. The transplantation of nuclei between two species of Xenopus. Dev Biol. 5: 68–83.
Halme DG and DA Kessler. 2006. FDA regulation of stem-cell-based therapies. N Engl J Med. 355: 1730–1735.
Han B, Y Lu, B Meng et al. 2009. Cellular loss after allogenic hepatocyte transplantation. Transplantation 87: 1–5.
Hanna JH, K Saha and R Jaenisch. 2010. Pluripotency and cellular reprogramming: facts, hypotheses, unresolved issues. Cell 143: 508–525.
Hay DC, J Fletcher, C Payne et al. 2008. Highly efficient differentiation of hESCs to functional hepatic endoderm requires ActivinA and Wnt3a signaling. Proc Natl Acad Sci USA. 105: 12301–12306.
Hay DC, S Pernagallo, JJ Diaz-Mochon et al. 2011. Unbiased screening of polymer libraries to define novel substrates for functional hepatocytes with inducible drug metabolism. Stem Cell Res. 6: 92–102.
Heo J, VM Factor, T Uren et al. 2006. Hepatic precursors derived from murine embryonic stem cells contribute to regeneration of injured liver. Hepatology. 44: 1478–1486.
Hockemeyer D, F Soldner, C Beard et al. 2009. Efficient targeting of expressed and silent genes in human ESCs and iPSCs using zinc-finger nucleases. Nat Biotechnol. 27: 851–857.
Hockemeyer D, F Soldner, EG Cook et al. 2008. A drug-inducible system for direct reprogramming of human somatic cells to pluripotency. Cell Stem Cell. 3: 346–353.
Holden C. 2005. Stem cell research. Korean cloner admits lying about oocyte donations. Science. 310: 1402–1403.
Hu BY, JP Weick, J Yu et al. 2010. Neural differentiation of human induced pluripotent stem cells follows developmental principles but with variable potency. Proc Natl Acad Sci USA. 107: 4335–4340.
Human Tissue and Cells Directive. Available http: //eur-lex.europa.eu/LexUriServ/LexUriServ?uri=OJ: L: 2004: 102: 0048: 0058: en: PDF Accessed on 4 March 2011.
Hyun I, O Lindvall, L Ahrlund-Richter et al. 2008. New ISSCR guidelines underscore major principles for responsible translational stem cell research. Cell Stem Cell. 3: 607–609.
Hyun I, P Taylor, G Testa et al. 2007. Ethical Standards for Human-to-Animal Chimera Experiments in Stem Cell Research. Cell Stem Cell. 1: 159–163.
Ieda M, JD Fu, P Delgado-Olguin et al. 2010. Direct reprogramming of fibroblasts into functional cardiomyocytes by defined factors. Cell. 142: 375–386.

Inamura M, K Kawabata, K Takayama et al. 2011. Efficient Generation of Hepatoblasts From Human ES Cells and iPS Cells by Transient Overexpression of Homeobox Gene HEX. Mol Ther. 19: 400–407.

International Society for Stem Cell Research. 2006. Guidelines for the conduct of human embryonic stem cell research. 15 p. Available: http: //www.isscr.org/guidelines/ISSCRhESCguidelines2006.pdf Accessed on 4 March 2011.

ISSCR Guidelines for the Clinical Translation of Stem Cells. 2008. http: //www.isscr.org/clinical_trans/ Accessed 4 March 2011.

Jaenisch R and R Young. 2008. Stem cells, the molecular circuitry of pluripotency and nuclear reprogramming. Cell. 132: 567–582.

Joffe S, MM Mello, EF Cook et al. 2007. Advance care planning in patients undergoing hematopoietic cell transplantation. Biol Blood Marrow Transplant. 13: 65–73.

Khetani SR and SN Bhatia. 2008. Microscale culture of human liver cells for drug development. Nat Biotechnol. 26: 120–126.

Kim D, CH Kim, JI Moon et al. 2009. Generation of human induced pluripotent stem cells by direct delivery of reprogramming proteins. Cell Stem Cell. 4: 472–476.

Kim K, A Doi, B Wen et al. 2010. Epigenetic memory in induced pluripotent stem cells. Nature 467: 285–290.

Klimanskaya I, Y Chung, S Becker et al. 2006. Human embryonic stem cell lines derived from single blastomeres. Nature. 444: 481–485.

Kneteman NM and DF Mercer . 2005. Mice with chimeric human livers: who says supermodels have to be tall? Hepatology. 41: 703–706.

Ko K and HR Scholer. 2006. Embryonic stem cells as a potential source of gametes. Semin Reprod Med 24: 322–329.

Lee G, EP Papapetrou, H Kim et al. 2009. Modelling pathogenesis and treatment of familial dysautonomia using patient-specific iPSCs. Nature. 461: 402–406.

Lengerke C, K Kim, O Lerou et al. 2007. Differentiation potential of histocompatible parthenogenetic embryonic stem cells. Ann N Y Acad Sci. 1106: 209–218.

Lensch MW, TM Schlaeger, LI Zon et al. 2007. Teratoma formation assays with human embryonic stem cells: a rationale for one type of human-animal chimera. Cell Stem Cell. 1: 253–258.

Lister R, M Pelizzola, YS Kida et al. 2011. Hotspots of aberrant epigenomic reprogramming in human induced pluripotent stem cells. Nature. 471: 68–73.

Lo B and L Parham . 2009. Ethical issues in stem cell research. Endocr Rev. 30: 204–213.

Locke, J.E., MJ Shamblott. and AM Cameron. 2009. Stem cells and the liver: clinical applications in transplantation. Advances in Surgery. 43: 35–51.

Lowrance WW and FS Collins. 2007. Ethics. Identifiability in genomic research. Science. 317: 600–602.

Lowry WE and WL Quan. 2010. Roadblocks en route to the clinical application of induced pluripotent stem cells. J Cell Sci. 123: 643–651.

Lyra AC, MB Soares, LF da Silva et al. 2007. Feasibility and safety of autologous bone marrow mononuclear cell transplantation in patients with advanced chronic liver disease. World J Gastroenterol. 3: 1067–1073.

Mack GS. 2011. ReNeuron and Stem Cells get green light for neural stem cell trials. News, Nature Biotechnology. 29: 95–97.

Maehr R, S Chen, M Snitow et al. 2009. Generation of pluripotent stem cells from patients with type 1 diabetes. Proc Natl Acad Sci USA. 106: 15768–15773.

Mateizel I, N De Temmerman, U, Ullmann et al. 2006. Derivation of human embryonic stem cell lines from embryos obtained after IVF and after PGD for monogenic disorders. Hum Reprod. 21: 503–511.

McGuire AL, T Caulfield and MK Cho. 2008. Research ethics and the challenge of whole-genome sequencing. Nat Rev Genet. 9: 152–156.

Medicinal Products for Human Use Directive (2003/63/EC, amending Directive 2001/83/EC). Available http: //ec.europa.eu/health/files/eudralex/vol-1/dir_2003_63/dir_2003_63_en.pdf Accessed on 4 March 2011.

Moriya K, M Yoshikawa, Y Ouji et al. 2008. Embryonic stem cells reduce liver fibrosis in CCl4-treated mice. Int J Exp Pathol. 89: 401–409.

Müller FJ, J Goldmann, P Löser et al. 2010. A call to standardize teratoma assays used to define human pluripotent cell lines. Cell Stem Cell. 6: 412–414.

Nakagawa M, M Koyanagi, K Tanabe et al. 2008. Generation of induced pluripotent stem cells without Myc from mouse and human fibroblasts. Nat Biotechnol. 26: 101–106.

National Commission for the Protection of Human Subjects of Biomedical and Behavioral Research. 1979. The Belmont Report: Ethical principles and guidelines for the protection of human subjects of biomedical and behavioral research. Available http: //ohsr.od.nih.gov/guidelines/belmont.html . Accessed 4 March 2011.

[NAS-U.S.]Amendments to the National Academies' guidelines for Human Embryonic Stem Cell Research. 2007. Available at: http: //www.nap.edu/openbook.php?record_id=11278&page=167.

[NRC-U.S] Guidelines for Human Embryonic Stem Cell Research. 2005. National Research Council (U.S.) and the Institute of Medicine, the Board on Life Sciences and the Board on Health Sciences Policy. National Academies Press. Washington DC. Available at: http: //www.nap.edu/openbook.php?isbn=0309096537

O'Malley J, K Woltjen and K Kaji. 2009. New strategies to generate induced pluripotent stem cells. Curr Opin Biotechnol. 20: 516–521.

Park IH, N Arora, H Huo et al. 2008. Disease-specific induced pluripotent stem cells. Cell. 134: 877–886.

Payne CM, K Samuel, A Pryde et al. 2011. Persistence of functional hepatocyte-like cells in immune-compromised mice. Liver Int. 31: 254–262.

Petersen B and T Shupe. 2008. Location is everything: the liver stem cell niche. Hepatology. 47: 1810–1812.

Petersen BE, WC Bowen, KD Patrene et al. 1999. Bone marrow as a potential source of hepatic oval cells. Science. 284: 1168–1170.

PHS (Public Health Service) Guideline on Infectious Disease Issues in Xenotransplantation Available http: //www.fda.gov/BiologicsBloodVaccines/GuidanceCompliance RegulatoryInformation/Guidances/Xenotransplantation/ucm074727.htm Accessed on 4 March 2011.

Piscaglia AC, M Campanale, A Gasbarrini et al. 2010. Stem cell-based therapies for liver diseases: state of the art and new perspectives. Stem Cells Int. 2010: 259461.

Polo JM, S Liu, ME Figueroa et al. 2010. Cell type of origin influences the molecular and functional properties of mouse induced pluripotent stem cells. Nat Biotechnol. 28: 848–855.

Raya A, I Rodríguez-Pizà, G Guenechea et al. 2009. Disease-corrected haematopoietic progenitors from Fanconi anaemia induced pluripotent stem cells. Nature. 460: 53–59.

Razonable RR, JY Findlay, A O'Riordan et al. 2011. Critical care issues in patients after liver transplantation. Liver Transpl. (in press)

Reflection paper on stem cell-based medicinal products EMA/CAT/571134/2009 Availablehttp: //www.ema.europa.eu/docs/en_GB/document_library/Scientific_guideline/2011/02/WC500101692.pdf Accessed on 4 March 2011.

Revazova ES, NA Turovets, OD Kochetkova et al. 2007. Patient-specific stem cell lines derived from parthenogenetic blastocysts. Cloning Stem Cells. 9: 432–449.

Robertson NJ, FA Brook, RL Gardner et al. 2007. Embryonic stem cell-derived tissues are immunogenic but their inherent immune privilege promotes the induction of tolerance. Proc Natl Acad Sci USA. 104: 20920–20925.

Ruhnke M, H Ungefroren, A Nussler et al. 2005. Differentiation of *in vitro*-modified human peripheral blood monocytes into hepatocyte-like and pancreatic islet-like cells. Gastroenterology. 128: 1774–1786.

Russo FP, MR Alison, BW Bigger et al. 2006. The bone marrow functionally contributes to liver fibrosis. Gastroenterology. 130: 1807–1821.
Rv Kelly and Lindsay [1998] EWCA Crim 1578 (14th May, 1998). The Human Act (UK 2004) http://www.legislation.gov.uk/ukpga/2004/30/pdfs/ukpga_20040030_en.pdf/ Accessed on 23 January 2012.
Saha K and R Jaenisch. 2009. Technical challenges in using human induced pluripotent stem cells to model disease. Cell Stem Cell. 5: 584–595.
Sancho-Bru P, M Najimi, M Caruso et al. 2009. Stem and progenitor cells for liver repopulation: can we standardise the process from bench to bedside? Gut. 58: 594–603.
Sato Y, H Araki, J Kato et al. 2005 .Human mesenchymal stem cells xenografted directly to rat liver are differentiated into human hepatocytes without fusion. Blood. 106: 756–763.
Seoane JA, DA Casado and A Rocha. 2008. Alternative consent models for biobank. Bioethics. 8: 440–447.
Si-Tayeb K, FK Noto, M Nagaoka et al. 2010. Highly efficient generation of human hepatocyte-like cells from induced pluripotent stem cells. Hepatology. 51: 297–305.
Soldner F, D Hockemeyer, C Beard et al. 2009. Parkinson's disease patient-derived induced pluripotent stem cells free of viral reprogramming factors. Cell. 136: 964–977.
Song Z, J Cai, Y Liu and et al.2009. Efficient generation of hepatocyte-like cells from human induced pluripotent stem cells. Cell Res. 19: 1233–1242.
Spence JR, CN Mayhew, SA Rankin et al. 2011. Directed differentiation of human pluripotent stem cells into intestinal tissue in vitro. Nature. 470: 105–109.
St John J and R Lovell-Badge. 2007. Human-animal cytoplasmic hybrid embryos, mitochondria, and an energetic debate. Nat Cell Biol. 9: 988–992.
St John JC and G Schatten. 2004. Paternal mitochondrial DNA transmission during nonhuman primate nuclear transfer. Genetics. 167: 897–905.
Stutchfield BM, SJ Forbes and SJ Wigmore. 2010. Prospects for stem cell transplantation in the treatment of hepatic disease. Liver Transpl. 16: 827–836.
Sullivan GJ, DC Hay, IH Park et al. 2010a. Generation of functional human hepatic endoderm from human induced pluripotent stem cells. Hepatology. 51: 329–335.
Sullivan GJ, Y Bai J Fletcher et al. 2010b. Induced pluripotent stem cells: epigenetic memories and practical implications. Mol Hum Reprod. 16: 880–885.
Takahashi K and S Yamanaka. 2006. Induction of pluripotent stem cells from mouse embryonic and adult fibroblast cultures by defined factors. Cell. 126: 663–676.
Terai S, T Ishikawa, K Omori et al. 2006. Improved liver function in patients with liver cirrhosis after autologous bone marrow cell infusion therapy. Stem Cells. 24: 2292–2298.
The Clinical Trials Directive (2001/20/EC). Available http: //ec.europa.eu/health/human-use/clinical-trials/index_en.htm Accessed on 4 March 2011.
The European Group on Ethics in Science and New Technologies to the European Commission. Opinion on the ethical aspects of patenting inventions involving human stem cells http://ec.europa.eu/bepa/european-group-ethics/docs/avis16_complet_en.pdf/Accessed on 23 January 2012.
The Food and Drug Administration. 1997. Proposed approach to regulation of cellular and tissue-based products. Available http: //www.fda.gov/downloads/BiologicsBloodVaccines/GuidanceComplianceRegulatoryInformation/Guidances/Tissue/UCM062601.pdf Accessed on 4 March 2011.
The Human Act (UK 2004) http://www.legislation.gov.uk/ukpga/2004/30/pdfs/ukpga_20040030_en.pdf/Accessed on 23 January 2012.
Thomson JA, J Itskovitz-Eldor, SS Shapiro et al. 1998. Embryonic stem cell lines derived from human blastocysts. Science. 282: 1145–1147.
Touboul T, NR Hannan, S Corbineau et al. 2010. Generation of functional hepatocytes from human embryonic stem cells under chemically defined conditions that recapitulate liver development. Hepatology. 51: 1754–1765.

Universal Declaration on Bioethics and Human Rights, UNESCO. 2005. Available: http: //www.unesco.org/new/en/social-and-human-sciences/themes/bioethics/bioethics-and-human-rights/Accessed on 30 March 2011.
US Food and Drug Administration. 2007. Good clinical practice in FDA-regulated clinical trials. Available: http: //www.fda.gov/ScienceResearch/SpecialTopics/RunningClinicalTrials/GuidancesInformationSheetsandNotices/default.htm Accessed 4 March 2011.
Verlinsky Y, J Cohen, S Munne et al. 2004. Over a decade of experience with preimplantation genetic diagnosis. Fertil Steril. 82: 302–303.
Vierbuchen T, A Ostermeier, ZP Pang et al. 2010. Direct conversion of fibroblasts to functional neurons by defined factors. Nature. 463: 1035–1041.
Weintraub H, SJ Tapscott, RL Davis et al. 1989. Activation of muscle-specific genes in pigment, nerve, fat, liver, and fibroblast cell lines by forced expression of MyoD. Proc Natl Acad Sci USA. 86: 5434–5438.
World Medical Association. 2004. Declaration of Helsinki. Available: http: //www.wma.net/en/30publications/10policies/b3/index.html. Accessed 4 March 2011.
Wu DC, AS Boyd and KJ Wood. 2007. Embryonic stem cell transplantation: potential applicability in cell replacement therapy and regenerative medicine. Frontiers in Bioscience. 12: 4525–4535.
Xia X, Y Zhang, CR Zieth et al. 2007. Transgenes delivered by lentiviral vector are suppressed in human embryonic stem cells in a promoter-dependent manner. Stem Cells Dev. 16: 167–176.
Xie H, M Ye, R Feng et al. 2004. Stepwise reprogramming of B cells into macrophages. Cell. 117: 663–676.
Yamamoto H, G Quinn, A Asari et al. 2003. Differentiation of embryonic stem cells into hepatocytes: biological functions and therapeutic application. Hepatology. 37: 983–893.
Ye Z, H Zhan, P Mali et al. 2009. Human-induced pluripotent stem cells from blood cells of healthy donors and patients with acquired blood disorders. Blood. 114: 5473–54780.
Zhao XY, W Li, Z Lv et al. 2009. iPS cells produce viable mice through tetraploid complementation. Nature. 461: 86–90.
Zhou H, LE Rogler, L Teperman et al. 2007. Identification of hepatocytic and bile ductular cell lineages and candidate stem cells in bipolar ductular reactions in cirrhotic human liver. Hepatology. 45: 716–724.
21 C.F.R. § 610.12 and 18. Available http: //www.accessdata.fda.gov/scripts/cdrh/cfdocs/cfcfr/CFRSearch.cfm?fr=610.12 Accessed on 4 March 2011.
42 USCA § 262. Available http: //www.selectagents.gov/resources/42USC262.pdf.

Index

Color Plate Section

Chapter 4

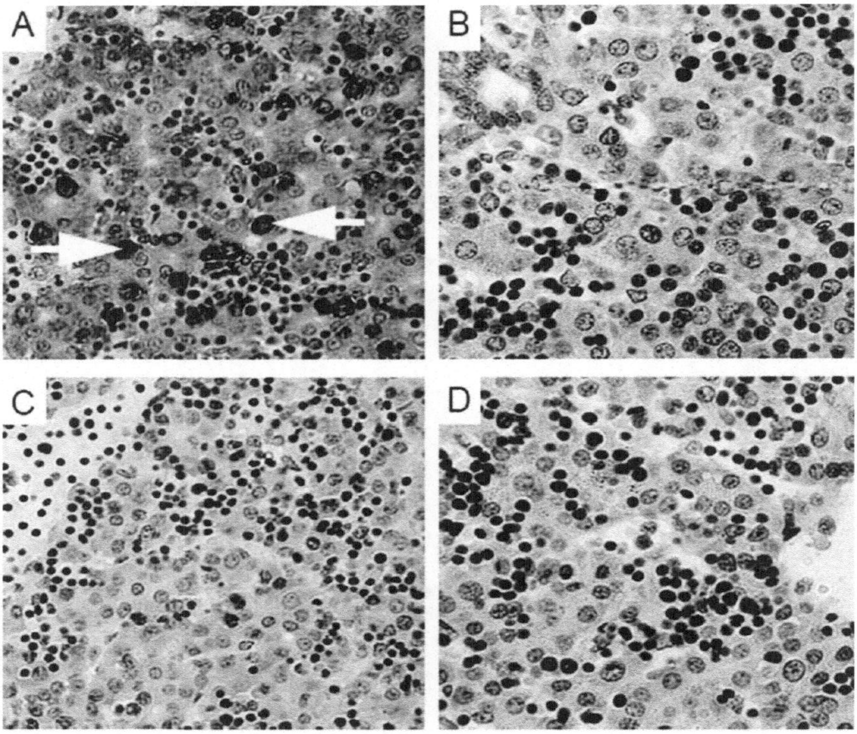

Figure 4.2. WIF-1 is an important factor in human liver development.
Wnt Inhibitory Factor-1 (WIF-1) immunostaining was carried out in human T1 (A) and T2 (B) embryonic liver paraffin embedded sections. WIF-1 expression was pronounced in areas adjacent to developing hepatoblasts in T1 liver (see white arrows, A). In contrast, WIF-1 expression was not detected in T2 liver. Isotype controls stained negative in T1 (C) and T2 (D) liver demonstrating specificity. Immunohistochemistry images were captured using a Leica FW4000 brightfield microscope. Images are at x20 magnification.

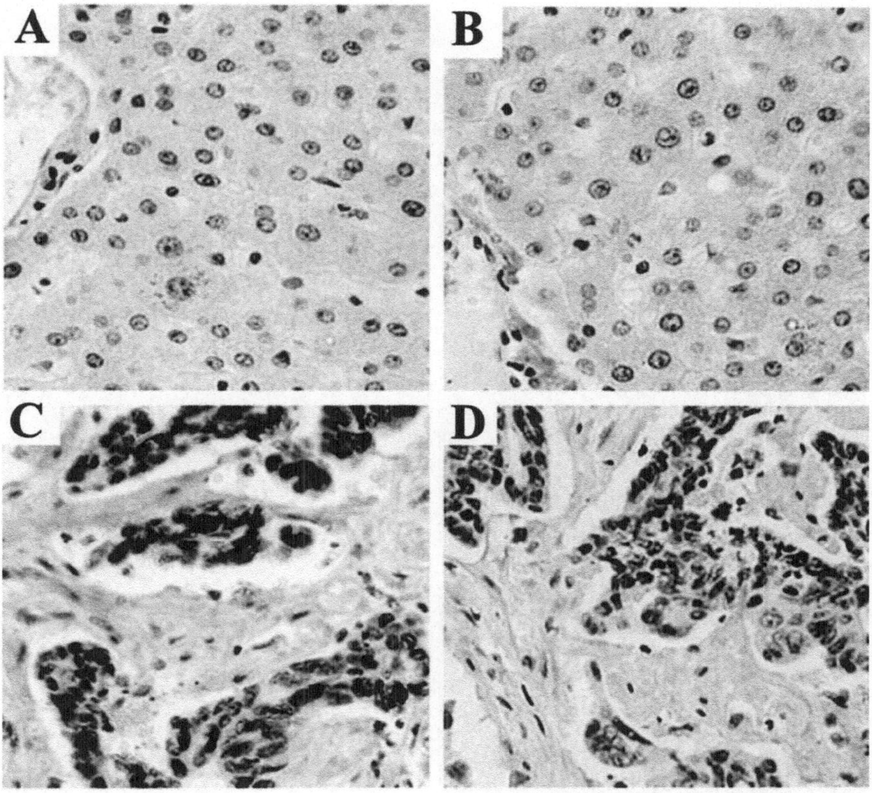

Figure 4.3. The absence of WIF-1 expression in adult human liver and its reactivation during cancer.

WIF-1 immunostaining in normal and diseased human adult liver. WIF-1 expression was not detected in normal adult liver parenchyma (A). In contrast WIF-1 expression was detected in adult liver with tumours containing SCM (C). Immunostaining controls stained negative, (B and D) thereby demonstrating specificity. Images were captured using a Leica FW4000 brightfield microscope at x20 magnification.

Chapter 6

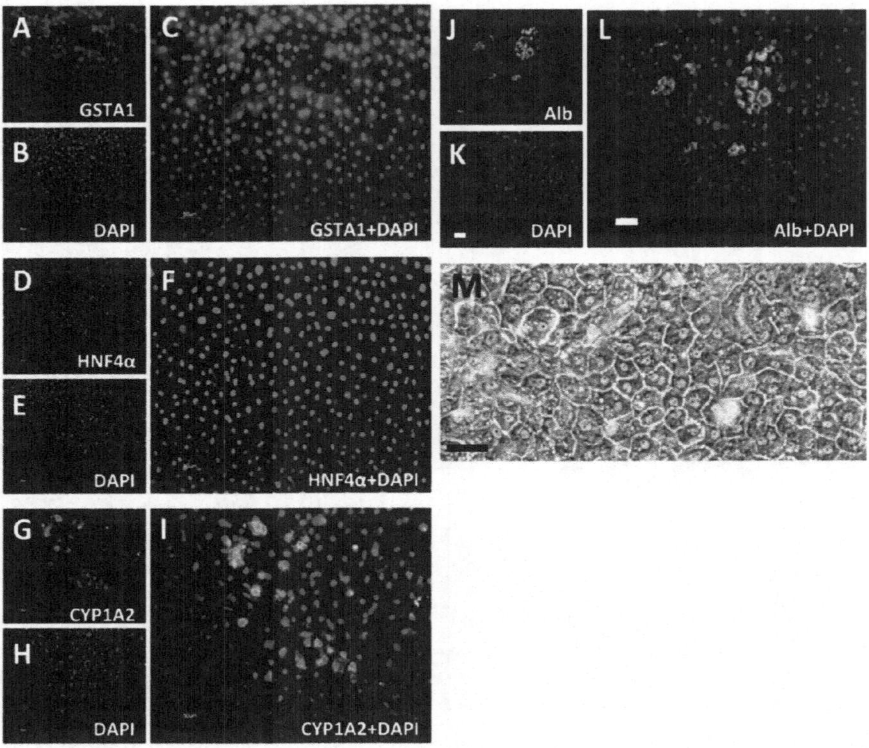

Figure 6.1. Microphotographs of functional human embryonic stem cells derived hepatocyte-like cells, cultured in a feeder-free system. A: Hepatocyte-like cells positive for Glutathione S-transferase 1 (GSTA1), a Phase II enzyme crucial for drug metabolism. B: Nuclei stained with DAPI. C: Merged photograph of A and B. D: HNF4α positive cells. The transcription factor HNF4α is important during differentiation to hepatocytes, and is used as a marker for differentiated hepatocytes. E: Nuclei stained with DAPI. F: Merged photograph of D and E. G: Cells positive for CYP1A2, a Phase I drug metabolizing enzyme and a marker of functional hepatocytes. H: Nuclei stained with DAPI. F: Merged photograph of G and H. J: Albumin staining, showing production of albumin in hepatocyte-like cells. K: Nuclei stained with DAPI. L: Merged photograph of J and K. M: Bright-field microscopic image displaying typical hepatocyte morphology of polygonal cells with bright nucleus and dark cytoplasm, large cytoplasm to nucleus ratio, nuclei with one to two prominent nucleoli, and numerous bi- or poly nucleated cells. Scale bar= 50μM. (Courtesy of Cellartis AB).

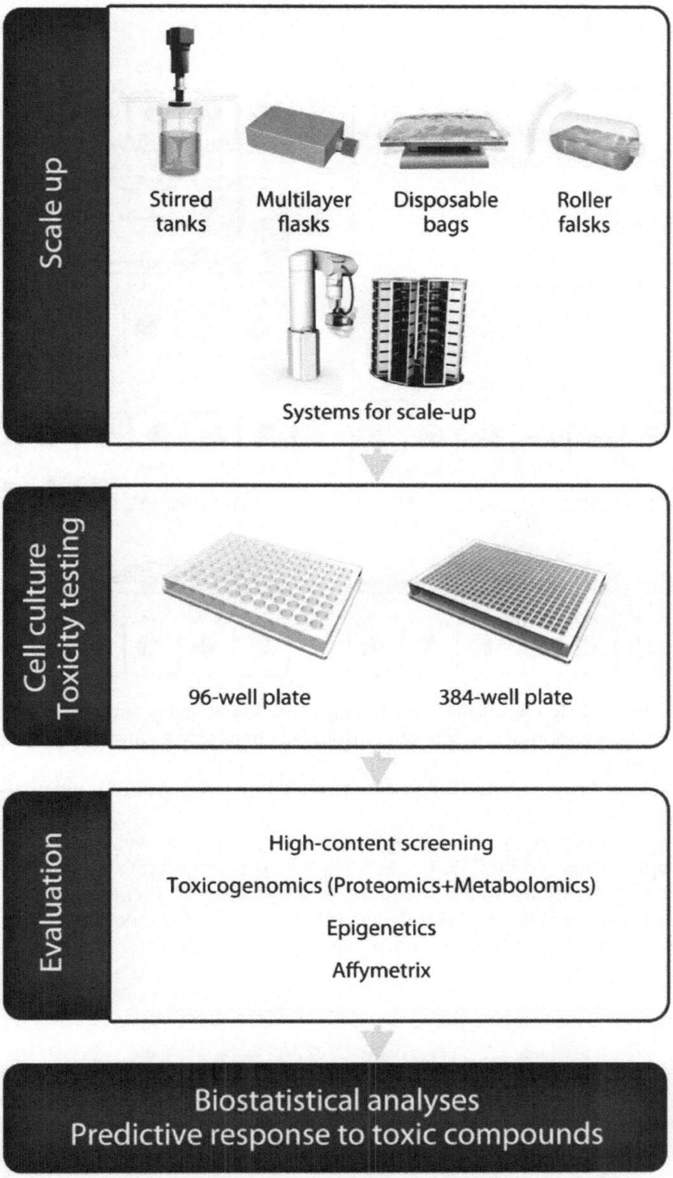

Figure 6.3. Illustration of steps essential for a high through-put hepatoxicity assay. First, the cell number should be possible to scale up, without loss of viability and pluripotency. Second, for cell cultures and toxicity testing in high through-put, the tests need to be performed in multiwell plates suitable for robot systems. Finally, to predict hepatocyte specific toxicity, evaluations with different methods and biostatistical analysis would benefit automated systems, such as high-content screening and affymetrix.

Chapter 9

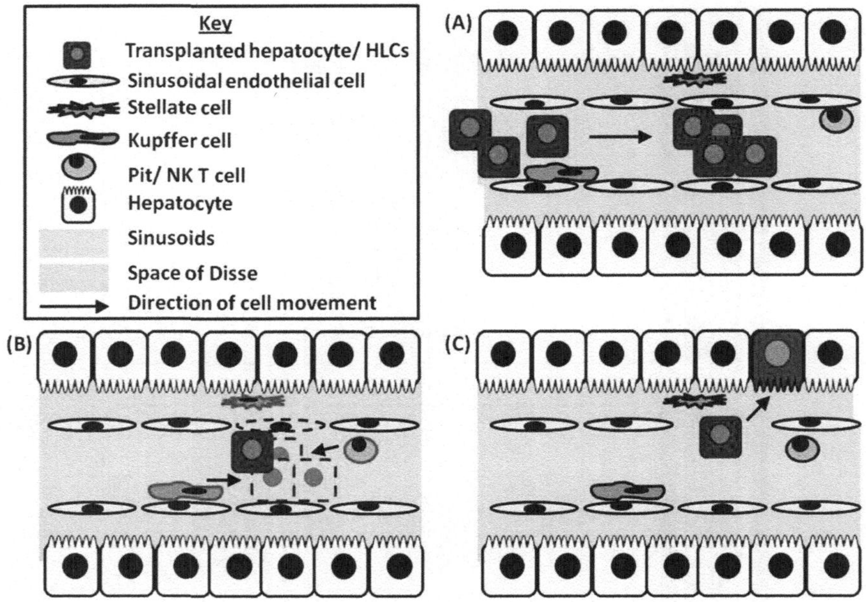

Figure 9.1. The route to engraftment for transplanted hepatocytes and hepatocyte-like cells. Schematic representation of the fate of transplanted hepatocytes and HLCs upon entering the liver sinusoids.

A. Transplanted hepatocytes/hepatocyte-like cells enter the sinusoids where they become lodged.
B. The majority of hepatocytes undergo anoikis and/or are removed as a result of clearance by Kupffer cells and Pit/NK T cells. Stellate cells become activated and release factors which may contribute to the disruption of sinusoidal endothelial barrier.
C. Following disruption of sinusiodal endothelial barrier, hepatocytes cross the Space of Disse and integrate into the parenchyma.